T0280488

Satellite Earth Observations and Their Impact on Society and Policy

Masami Onoda · Oran R. Young
Editors

Satellite Earth Observations and Their Impact on Society and Policy

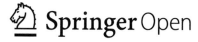

Editors
Masami Onoda
Japan Aerospace Exploration Agency
Tokyo
Japan

Oran R. Young
Bren School of Environmental Science and
  Management
University of California
Santa Barbara, CA
USA

ISBN 978-981-10-9949-6   ISBN 978-981-10-3713-9   (eBook)
DOI 10.1007/978-981-10-3713-9

Printed on acid-free paper

This Springer imprint is published by Springer Nature
The registered company is Springer Nature Singapore Pte Ltd.
The registered company address is: 152 Beach Road, #21-01/04 Gateway East, Singapore 189721, Singapore

To Molly Macauley, an inspirational contributor to the project, whose expertise in space policy and economics will always provide invaluable guidance to the space community.

Foreword

As gravitationally bound beings, humanity has long dreamed of soaring into the atmosphere and beyond to gaze down upon the Earth, our home, both out of curiosity and an instinctive desire to better know our place in the world.

Thanks to human ingenuity, we have realized this possibility. While only a select few have had the privilege to experience this unique viewpoint first-hand, Earth observation satellites have opened all of our eyes and, in the process, made us acutely aware of the importance of our planet's environment. Geostationary satellites in particular—in addition to supporting telecommunications, positioning, broadcasting, and surveillance—provide unique and important insights into the dynamics of the Earth's environment and climate.

While these benefits are well recognized, Earth observations also provide intangible and philosophical benefits. Their views remind us that we are all global citizens of planet Earth, living together in a fragile ecosystem that requires our utmost care and respect, urging international peace and cooperation.

Earth observations are usually discussed in the context of science, but in this volume we explore the political aspects, driven by a desire to understand the interplay between environmental policy and technical developments in Earth observation. This volume covers findings first presented at a November 2015 workshop held in Tokyo, Japan, on "Assessing the Impact of Satellite Earth Observation on Society and Policy." The contents explore policy-relevant satellite observations in a number of key areas, as well as the coordination and governance of Earth-observing systems required to maximize their contribution to solving a range of environmental problems.

Negotiation and compromise are necessary parts of environmental policy. Earth observations are fundamental instruments that facilitate discourse by providing a mutual understanding of our environmental issues on a global scale.

I sincerely hope the findings herein accelerate our discussion and action.

Akimasa Sumi
President of the National Institute
for Environmental Studies
Japan
Professor Emeritus
The University of Tokyo

Advisory Board and Project Members

International Advisory Board Workshop
Assessing the Impact of Satellite Earth Observation on Society and Policy
9–10 November 2015
Tokyo, Japan

Summary of Conclusions

An International Advisory Board recently met in Tokyo with the overarching goal of *assessing the impact of satellite earth observation on society and policy.*[1] A series of roundtable discussions invited perspectives from experts to understand how Earth observations contribute to environmental and other policymaking, and how space agencies establish links between their programs, scientific advances, industrial innovation, and societal well-being. At the closing of the workshop, the Advisory Board concluded with the following key findings from the two days of meetings:

1. **Earth observations provide a unique window and perspective on our world, serving the betterment of all humankind by supporting policies aimed at sustainably managing natural and societal resources on an ever more populous, affluent, and interconnected planet Earth.**

   – For example, Earth observations can make an important contribution to addressing some of the world's greatest health risks including air pollution, water contamination, lack of sanitation, and risks related to increasing urbanization.

---

[1]The Advisory Board was convened in support of the project "Study on Methods for Assessing the Impact of Satellite Observations on Environmental Policy", funded by the Japan Science and Technology Agency (JST), and jointly carried out by the National Institute of Information and Communications Technologies (NICT), Keio University, Institute for Global Environmental Strategies (IGES), and the Japan Aerospace Exploration Agency (JAXA).

2. **Earth observations should be regarded as critical societal infrastructure. There is strong evidence that publicly open Earth observations are making positive, cost-effective contributions to solving a variety of high-priority environmental and societal problems.**

   – Studies on the socioeconomic benefits of improved global Earth observation systems show that the benefits outweigh the costs by orders of magnitude when subject to a free and open data policy. The European Copernicus program, for instance, is expected to return benefits to taxpayers valued 10 times higher than the costs.

3. **There is a need to develop appropriate institutions in the field of Earth observation through a process to ensure that the observations and prediction systems are comprehensively exploited for policymaking with full engagement of all stakeholders and end-users.**

   – The U.S. (Decadal Survey) and European (Copernicus) experiences provide fine examples of the benefits of "all-of-government" processes in defining satellite missions. At the global level, initiatives for greenhouse gases, forests, and other areas are being developed to support contributions to policy.

4. **Japan, together with its international partners, should identify and fill emerging gaps in next-generation space missions to guarantee full realization of all societal benefits of Earth observations derived from long-term continuity.**

   – The lack of a systematic, long-term plan for satellite environmental observation missions by the Japanese Space Plan is of particular concern.

5. **There is a changing paradigm for Earth observations, with non-governmental groups launching satellites, and with the growing popularity of small satellites, drones and crowd-sourcing/citizen science campaigns, which are associated with the rapid development of data technology and applications.**

   – The increasing number of rapid-response, cost-effective and high-performance satellite missions, together with the possibility of exploiting "big-data", provides opportunities, as well as challenges, for enhanced global Earth observations.

List of Participants

- International Advisory Board members:

  Oran Young (Chair), University of California, Santa Barbara
  Josef Aschbacher, European Space Agency (ESA)
  Carlos Dora, World Health Organization (WHO)
  Jinlong Fan, China Meteorological Administration (CMA)
  Claire Jolly, Organisation for Economic Co-operation and Development
  (OECD)
  Murielle Lafaye, French National Centre for Space Studies (CNES)
  Molly Macauley, Resources for the Future (RFF)
  Teruyuki Nakajima, Japan Aerospace Exploration Agency (JAXA)
  Michael Obersteiner, International Institute for Applied Systems Analysis
  (IIASA)
  Ake Rosenqvist, soloEO
  Sir Martin Sweeting, Surrey Satellite Technology Ltd. (SSTL)
  Tatsuya Yokota, National Institute for Environmental Studies (NIES)

- Japan Science and Technology Agency (JST) Strategic Basic Research
  Programs (Research Institute of Science and Technology for Society) "Study on
  Methods for Assessing the Impact of Satellite Observations on Environmental
  Policy" Project members:

  Yasuko Kasai, National Institute of Information and Communications
  Technologies (NICT)
  Akiko Aizawa, National Institute of Informatics (NII)
  Setsuko Aoki, Keio University
  Akiko Okamatsu, Hosei University
  Masami Onoda, Japan Aerospace Exploration Agency (JAXA)
  Henry Scheyvens, Institute for Global Environmental Strategies (IGES)
  and others

Special Collaboration

This book was published by the special collaboration of Dr. Yasuko Kasai of the National Institute of Information and Communications Technology (NICT), as Project Leader of the research titled "A Study on Methods for Objective/Quantitative Assessment of the Impact of Satellite Observations on Environmental Policy" under the program "Science of Science, Technology and Innovation Policy" of the Research Institute of Science and Technology for Society (RISTEX), Japan Science and Technology Agency (JST).

Acknowledgements

We are deeply grateful to the R&D program "Science of Science, Technology and Innovation Policy" of the Research Institute of Science and Technology for Society (RISTEX), Japan Science and Technology Agency (JST) for the financial support that made this project possible. We would like to thank the program's advisors, Prof. Yoshifumi Yasuoka, Dr. Akio Morita, and Ms. Michiko Igarashi, for their valuable guidance and Ms. Shiho Hamada and Ms. Azusa Sano of JST for their support throughout the project. We acknowledge with gratitude the National Institute of Information and Communications Technology (NICT), particularly former President Masao Sakauchi and Vice President Fumihiko Tomita; the Institute for Global Environmental Strategies (IGES), especially Prof. Hironori Hamanaka; Keio University; Japan Aerospace Exploration Agency (JAXA), especially Mr. Shizuo Yamamoto and Ms. Akiko Suzuki; Hosei University; and the National Institute for Informatics (NII). We also thank Mr. Toru Fukuda for his support in initiating the project. Our deep gratitude goes to Prof. Akimasa Sumi of the National Institute of Environmental Studies (NIES) and all members of the International Advisory Board for their contributions to the project. We offer our special thanks to Matthew Steventon, Stephen Ward, and Jenny Harry for their outstanding work in editing the manuscript, and to Mr. Yosuke Nishida and Ms. Risa Takizawa of Springer Japan for overseeing the production of this volume.

Contents

Editors and Contributors

About the Editors

Masami Onoda is currently the U.S. and multilateral relations interface at the International Relations and Research Department of the Japan Aerospace Exploration Agency (JAXA). As an academic, she is fellow of the Institute of Global Environmental Strategies (IGES) and she is also engaged in the private sector as an advisor to the Singapore-based space debris start-up Astroscale Pte. Ltd. since its foundation in 2013. From 2009 to 2012, Dr. Onoda was a scientific and technical officer at the intergovernmental Group on Earth Observations (GEO) Secretariat in Geneva, Switzerland. From 2003 to 2008, while pursuing her graduate studies, she was invited to the JAXA Kansai Satellite Office in Higashiosaka as a space technology coordinator to support technology transfer to SMEs for the small satellite project SOHLA-1. From 1999 to 2003, she worked in the field of Earth observations at JAXA (then NASDA), serving on the Secretariat of the Committee on Earth Observation Satellites (CEOS). In 1999, she was seconded to the UN Office for Outer Space Affairs (UNOOSA) for the organization of the UNISPACE III conference. She holds a Ph.D. in global environmental studies (2009) and a master's degree in environmental management (2005), both from the Kyoto University Graduate School of Global Environmental Studies. Her undergraduate degree is in international relations from The University of Tokyo.

Oran R. Young is a renowned Arctic expert and a world leader in the fields of international governance and environmental institutions. His scientific work encompasses both basic research focusing on collective choice and social institutions, and applied research dealing with issues pertaining to international environmental governance and the Arctic as an international region. Professor Young served for 6 years as vice-president of the International Arctic Science Committee and was the founding chair of the Committee on the Human Dimensions of Global Change within the National Academy of Sciences in the U.S.A. He has chaired the Scientific Committee of the International Human Dimensions Programme on Global Environmental Change and the Steering Committee of the Arctic Governance Project.

Contributors

Josef Aschbacher Earth Observation Programmes, European Space Agency, Frascati (RM), Italy

Akiko Aizawa National Institute of Informatics (NII), Chiyoda-Ku, Tokyo, Japan

Setsuko Aoki Keio University Law School, Minato-Ku, Tokyo, Japan

Juraj Balkovič International Institute for Applied Systems Analysis, Laxenburg, Austria

Jetske A. Bouma International Institute for Applied Systems Analysis, Laxenburg, Austria

Hannes Böttcher International Institute for Applied Systems Analysis, Laxenburg, Austria

Jinlong Fan National Satellite Meteorological Center, China Meteorological Administration, Beijing, China

Lawrence Friedl NASA Applied Sciences Program, Earth Science Division, Science Mission Directorate, NASA Headquarters, Washington, D.C., USA

Steffen Fritz International Institute for Applied Systems Analysis, Laxenburg, Austria

Sabina Fuss International Institute for Applied Systems Analysis, Laxenburg, Austria

Peter Havlik International Institute for Applied Systems Analysis, Laxenburg, Austria

Christine Heumesser International Institute for Applied Systems Analysis, Laxenburg, Austria

Stefan Hochrainer International Institute for Applied Systems Analysis, Laxenburg, Austria

Kerstin Jantke International Institute for Applied Systems Analysis, Laxenburg, Austria

Brian Alan Johnson Natural Resources and Ecosystem Services, Institute for Global Environmental Strategies, Hayama, Kanagawa, Japan

Claire Jolly Organisation for Economic Co-operation and Development, Paris Cedex 16, France

Yasuko Kasai National Institute of Information and Communications Technology (NICT), Koganei, Tokyo, Japan

Nikolay Khabarov International Institute for Applied Systems Analysis, Laxenburg, Austria

Barbara Koch International Institute for Applied Systems Analysis, Laxenburg, Austria

Florian Kraxner International Institute for Applied Systems Analysis, Laxenburg, Austria

Onno J. Kuik International Institute for Applied Systems Analysis, Laxenburg, Austria

Murielle Lafaye Prospective Spatiale et Enjeux Socio-économiques, CNES – DIA/IP - Bpi 2903, Toulouse Cedex 9, France

Sylvain Leduc International Institute for Applied Systems Analysis, Laxenburg, Austria

Junguo Liu International Institute for Applied Systems Analysis, Laxenburg, Austria

Wolfgang Lucht International Institute for Applied Systems Analysis, Laxenburg, Austria

Molly Macauley (Deceased) Resources for the Future (RFF), Washington, DC, USA

Ian McCallum International Institute for Applied Systems Analysis, Laxenburg, Austria

Reinhard Mechler International Institute for Applied Systems Analysis, Laxenburg, Austria

Elena Moltchanova International Institute for Applied Systems Analysis, Laxenburg, Austria

Teruyuki Nakajima Earth Observation Research Center (EORC), Japan Aerospace Exploration Agency (JAXA), Tsukuba, Ibaraki, Japan

Michael Obersteiner International Institute for Applied Systems Analysis, Laxenburg, Austria

Osamu Ochiai Group on Earth Observations, Geneva 2, Switzerland

Akiko Okamatsu Faculty of Sustainability Studies, Hosei University, Chiyoda, Tokyo, Japan

Masami Onoda International Relations and Research Department, Japan Aerospace Exploration Agency, Tokyo, Japan

Belinda Reyers International Institute for Applied Systems Analysis, Laxenburg, Austria

Ake Rosenqvist solo Earth Observation (soloEO), Tokyo, Japan

Barbara Ryan Group on Earth Observations, Geneva 2, Switzerland

Felicjan Rydzak International Institute for Applied Systems Analysis, Laxenburg, Austria

Hiromitsu Samejima Natural Resources and Ecosystem Services, Institute for Global Environmental Strategies, Kanagawa, Japan

Tomohiro Sato Big Data Analytics Laboratory, Big Data Integration Research Center, National Institute of Information and Communications Technology (NICT), Koganei, Tokyo, Japan

Henry Scheyvens Natural Resources and Ecosystem Services, Institute for Global Environmental Strategies, Kanagawa, Japan

Christian Schill International Institute for Applied Systems Analysis, Laxenburg, Austria

Christine Schleupner International Institute for Applied Systems Analysis, Laxenburg, Austria

Erwin Schmid International Institute for Applied Systems Analysis, Laxenburg, Austria

Uwe A. Schneider International Institute for Applied Systems Analysis, Laxenburg, Austria

Robert J. Scholes International Institute for Applied Systems Analysis, Laxenburg, Austria

Linda See International Institute for Applied Systems Analysis, Laxenburg, Austria

Rastislav Skalský International Institute for Applied Systems Analysis, Laxenburg, Austria

Alexey Smirnov International Institute for Applied Systems Analysis, Laxenburg, Austria

Olav Schram Stokke Department of Political Science, University of Oslo, Oslo, Norway; Fridtjof Nansen Institute, Lysaker, Norway

Martin Sweeting Surrey Satellite Technology Limited, Guildford, UK

Jana Szolgayova International Institute for Applied Systems Analysis, Laxenburg, Austria

Zuzana Tarasovičová International Institute for Applied Systems Analysis, Laxenburg, Austria

Hong Yang International Institute for Applied Systems Analysis, Laxenburg, Austria

Tatsuya Yokota Center for Global Environmental Research, National Institute for Environmental Studies (NIES), Tsukuba, Ibaraki, Japan

Oran R. Young Bren School of Environmental Science and Management, University of California, Santa Barbara, CA, USA

Part I
Background and Introduction

Chapter 1
Satellite Earth Observations in Environmental Problem-Solving

Oran R. Young and Masami Onoda

1.1 Introduction

Large-scale environmental problems have become prominent byproducts of the interactions between human activities and the biophysical settings in which they occur. Some of these problems, such as the destruction of tropical forests, arise in areas within the jurisdiction of individual states but have consequences (e.g., biodiversity loss, climate change) that are regional to global in scope. Other problems, including the seasonal thinning of the stratospheric ozone layer, occur in areas that lie beyond the jurisdiction of nation states. Still others, like marine dead zones, originate in areas within the jurisdiction of individual states but can spread and have impacts on areas outside these jurisdictional boundaries. In all these cases, efforts to come to terms with the relevant problems require international (often global) responses that call for transboundary cooperation and, more often than not, eventuate in the development of regulatory arrangements designed to limit or even prohibit the human actions that give rise to the problems.

In recent decades, international environmental regimes have arisen to address problems of this sort. These institutional arrangements provide governance systems designed to address a variety of issues ranging from atmospheric problems like acid rain or the seasonal thinning of the stratospheric ozone layer, marine problems like

O.R. Young (✉)
Bren School of Environmental Science and Management, University of California,
Santa Barbara, CA 93106-5131, USA
e-mail: oran.young@gmail.com

M. Onoda
International Relations and Research Department, Japan Aerospace Exploration Agency,
Ochanomizu Sola City, 4-6, Kanda Surugadai, Chiyoda-Ku, Tokyo 101-8008, Japan
e-mail: onoda.masami@jaxa.jp

© The Author(s) 2017
M. Onoda and O.R. Young (eds.), *Satellite Earth Observations and Their Impact on Society and Policy*, DOI 10.1007/978-981-10-3713-9_1

3

oil pollution from offshore platforms and tankers, to terrestrial problems like desertification in drought-stricken areas or tropical deforestation.[1] Regimes have met with varying degrees of success, attributable to many factors. But in all cases, a necessary condition for making progress is the acquisition of the data and information needed to grasp the fundamental character of the problems themselves and to support the activities of those responsible for administering the provisions of the resultant regimes.

One increasingly important source of data for addressing environmental problems is satellite Earth observation. In many cases (e.g., for monitoring global ozone depletion or tropical deforestation), there is simply no alternative for consistently tracking changes in environmental conditions at a global scale, frequent time interval, and fine spatial resolution. For example, while it is possible in principle to track deforestation in an area like the Amazon Basin from the ground, satellite Earth observation data provide a far more effective and efficient way to acquire an accurate picture of the scope of the problem.

Earth observations have shown great potential for generating global scientific information related to various environmental issues, but they have not yet been fully incorporated as a measurement tool in the monitoring processes of environmental regimes. Despite the call for more data and information for decision-making, a considerable disconnect remains between the policy needs and the data and information supply from satellite observations. To help close this gap, it will be important to identify the roles of satellite Earth observations and to consider how to manage the activities involved in acquiring and applying data, in order to enhance the contributions of these observations to solving problems.

The main goal of this book is to investigate methodologies for assessing the roles and impacts of satellite Earth observations in addressing environmental problems and, in the process, to contribute to thinking about broader questions relating to the interfaces among science, technology, policy, and society. We consider the development of Earth observation systems and the efforts being made to link societal needs and global policy demands to them in China, Europe, Japan, and the United States, as well as the prospects for cooperation among key agencies in each of these states (or consortia of states in the case of Europe). The book also attempts to identify generic models that might prompt governments and organizations to consider the connection of science and technology—in particular satellite Earth observation—to society and policy. We assess existing arrangements (e.g., the Group on Earth Observations, GEO) for coordinating the efforts of the producers of Earth observation data (e.g., national space or environmental agencies) and the users of the data (e.g., government agencies, researchers, NGOs, and businesses). In addition, we consider whether there is a need to create a more highly developed governance system covering satellite observations that would deal with interactions between producers and users of satellite observations, and to develop a system of rules and procedures applicable to the activities of all those active in this realm.

---

[1]For a range of perspectives on international environmental regimes, see Young et al. (2008).

1.2   The Nature and Scope of Satellite Earth Observations[2]

Remote sensing is a relatively new term that arose in the 1950s–1960s in the U.S.[3] It refers to "the acquisition of information about an object without physical contact" (Colwell 1983). Humans employ remote sensing all the time, as our eyes detect the light reflected by objects beyond our reach. Earth observation by remote sensing involves the measurement of electromagnetic energy reflected from or emitted by the Earth's surface (or atmosphere), together with the establishment of relationships between the remotely sensed measurements and some phenomena of interest on the Earth's surface (or within the atmosphere) (Mather 1999). The terms *Earth observation* and *remote sensing* are used interchangeably in most parts of this book. While remote sensing encompasses the technical methods and the technology, the term Earth observation places emphasis on the purpose of the activity. The principal concern of the chapter is with the use of remote sensing, but we also include methods of global monitoring that employ technologies other than remote sensing, such as satellite navigations and communications (e.g., Automatic Identification Systems (AIS) of vessels).

The electromagnetic wavelengths typically utilized for remote sensing of the Earth's surface/atmosphere range from the ultraviolet to microwave regions of the spectrum (Fig. 1.1). Useful information can be extracted from remote-sensing data because different types of objects on the Earth's surface (or in the atmosphere) tend to have different electromagnetic reflectance/emittance properties. That is to say, they have different spectral characteristics. It is possible to identify many types of objects based on their spectral signature, i.e., their unique reflectance/emittance properties in some specific regions of the electromagnetic spectrum. In order to measure these signatures, satellites carry scanners and Charged Coupled Devices (CCDs) that collect digital data and transmit it by radio back to Earth. Because the data recorded by these sensors are digital, they can be manipulated in various ways to derive useful information from the spectral measurements.

Remote-sensing data from different parts of the electromagnetic spectrum convey different types of information about the Earth's land/water surface or atmosphere. Images acquired in the visible to intermediate, or shortwave, infrared wavelengths (typically called "optical" remote-sensing data) are often used for the monitoring of vegetation and other types of objects on the ground or atmosphere (e.g., clouds) that have distinctive colors. Multispectral optical sensors, which measure reflected solar radiation over a few (1 to approx. 30) relatively broad

---

[2]For further technical introductions to remote sensing and Earth observation technology, see Short (2008), Colwell (1983), Mather (1999).

[3]Short (2008) indicates that the term "remote sensing" had been coined in the mid-1950s by Ms. Evelyn Pruitt, a geographer/oceanographer with the U.S. Office of Naval Research (ONR), to take into account the new views from space obtained by the early meteorological satellites that were obviously more "remote" from their targets than the airplanes that up until then provided mainly aerial photos as the medium for recording images of the Earth's surface.

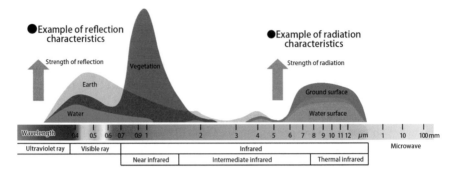

Fig. 1.1 Strength of reflection and radiation of electromagnetic waves from plants, earth, and water in each wavelength (JAXA 2016a, b, c, d)

portions of the optical spectrum, can differentiate between some types of objects on the surface/atmosphere. Hyperspectral optical sensors, which take measurements over many more (approx. 30 to several hundred) narrower portions of the spectrum, can discriminate between even more types of objects, but hyperspectral satellite imagery is still not yet widely available globally. Images acquired in the thermal infrared spectrum can provide information on the surface temperatures of objects and are often used to monitor the health of crops (e.g., leaf water content) or detect forest fires. Images acquired in different parts of the microwave spectrum can be used to monitor various vegetation/soil/water parameters. For example, short wavelength microwaves emitted from an active microwave satellite (e.g., a Synthetic Aperture Radar (SAR) satellite) are easily scattered by top surface materials (e.g., plant leaves in vegetated areas or the top of the soil layer in bare areas), providing useful information related to these materials (e.g., plant Leaf Area Index (LAI) or top soil moisture). Longer wavelength microwaves emitted by active microwave satellites can penetrate through these surfaces and provide other types of useful information (e.g., plant stem volume or subsurface soil moisture). Analysis of SAR images acquired at different time periods also allows for the measurement of surface deformations (e.g., land subsidence/uplift) or changes in water levels. Unlike visible and infrared satellite sensors, active microwave sensors are able to acquire information on the Earth's surface regardless of cloud cover. Passive microwave sensors measure radiation emitted by the Earth's surface at microwave wavelengths, and can be used to monitor soil moisture, snow melt, sea ice, sea surface temperatures, temperature profiles of the atmosphere, water vapor, ozone distribution, and precipitation. Finally, light detection and ranging (LIDAR) satellite data involve transmission of coherent laser light from the sensor and ranging to determine the height of objects on the Earth's surface (e.g., ground elevation, water depth, vegetation/building heights) or the abundance of different particles in the atmosphere, depending on the wavelength of the emitted laser. Other types of remote-sensing data for Earth observation exist, but the contributions in this book focus mainly on the types of data described here.

Other specifications of satellite remote-sensing instruments that have important implications for their environmental monitoring capabilities are the spatial and temporal resolutions at which they image the Earth. As might be expected, the spatial resolution of remote-sensing data describes the level of spatial detail of the data. The spatial resolution of the imagery generally determines the minimum size of objects that can be detected by the sensor. For example, an object that is at least 1 m wide and 1 m long can be separated (resolved) theoretically from its surrounding objects in an image with a spatial resolution of 1 m or finer. Commercial satellites today can produce images with very high (i.e., fine) spatial resolutions, 0.5 m or better. The temporal resolution of remote-sensing data describes the time interval at which images are acquired over the same location. Some satellites, typically those that acquire lower spatial resolution images (approx. 250 m or coarser, e.g., Terra/Aqua MODIS, Himawari-8) can image the entire globe daily, while other satellites that acquire higher spatial resolution images (approx. 30 m or finer, e.g., Landsat 8, Sentinel-2A) may only collect images of the same location every few weeks.

In summary, Earth observation by satellite remote sensing provides a powerful tool for monitoring various features of the Earth's environment not only because of the wide range of measurements that can be performed, but also because of the high frequency and large area for which data can be acquired. The digital era has made it possible to perform, store, analyze, and share data globally in a way that no one could imagine half a century ago. Available technology allows Earth observation data to be disseminated widely, and the need is stronger than ever to devise ways to put the data to work in meeting societal challenges.

1.3   Applications of Satellite Earth Observations

As a point of entry into the analysis of the links between Earth observations and public policy, we provide some examples dealing with atmospheric CO_2 monitoring, marine observations, and forest observations. Later chapters include more detailed accounts of these cases and link them to specific policy issues.

1.3.1   Atmospheric CO_2

As world leaders and the climate community met in Paris, France, for the 21st session of the UN Framework Convention on Climate Change (UNFCCC) Conference of the Parties (COP) in December 2015, global whole-atmospheric monthly mean CO_2 concentration observed by analysis of Japan's Greenhouse gases Observing SATellite "IBUKI" (GOSAT) was passing 400 ppm for the first time since GOSAT's launch in 2009 (Fig. 1.2). If this upward trend continues, further analysis is expected to show that the trend line of global CO_2 (indicating the

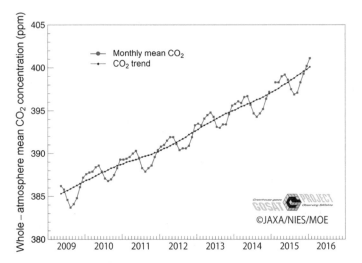

Fig. 1.2 Whole-atmosphere monthly mean CO_2 concentration derived from GOSAT observations. The CO_2 trend is calculated by removing averaged seasonal fluctuations from the monthly CO_2 time series (JAXA/NIES/MOE 2016)

trend after removing seasonal variations) exceeded 400 ppm around March 2016, which will mean that current global atmospheric CO_2 concentrations substantially exceed 400 ppm (JAXA/NIES/MOE 2016; JAXA 2016a, b, c, d).

Anticipating the policy needs of their governments, space agencies have been developing missions in recent years to provide detailed global measurements of various greenhouse gases. GOSAT, the first satellite designed specifically to monitor greenhouse gases from space, provides critical information on atmospheric CO_2 trends, as it is able to provide global, column-averaged CO_2 concentrations every three days, and has been monitoring long-term trends since its launch in 2009. In addition to GOSAT, NASA's Orbiting Carbon Observatory-2 (OCO-2) has been collecting data operationally since September 2014. OCO-2 is the first NASA mission designed to collect space-based measurements of atmospheric CO_2 with the precision, resolution, and coverage needed to characterize the processes controlling its accumulation in the atmosphere. OCO-2 is able to generate a global product every 16-days showing column-averaged concentrations of CO_2 in the order of a few kilometers in resolution. In combination with chemical transport modeling and image compositing techniques, highly detailed global maps have been developed which clearly show periodic daily fluctuations of atmospheric CO_2 over key sources and sinks (Fig. 1.3). In the future, it is possible that near-real-time monitoring of CO_2 from space may be available to support all phases of environmental policy, including definition, monitoring, and enforcement.

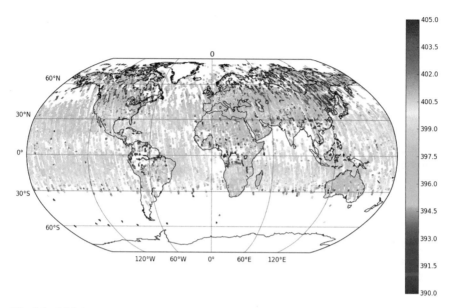

Fig. 1.3 OCO-2 Level 2 Lite product. One month CO_2 column-averaged dry air mole fraction (OCO-2 Science Team/Michael Gunson and Annmarie Eldering 2015)

1.3.2 Marine Pollution

1.3.2.1 Operational Oil Pollution Monitoring

The environmental impacts of large-scale industrial oil spills are quite well known. But smaller day-to-day leaks and discharges from passing ships also have a profound effect on marine ecosystems. Since 1998, Kongsberg Satellite Services (KSAT) in Tromsø, Norway, has operated a near-real-time oil monitoring system that relies primarily on SAR data, which is well suited to detecting surface slicks and can be used in all weather conditions. Missions such as Envisat (past), Sentinel-1, Radarsat-2, TerraSAR-X, and COSMO-SkyMed are used in combination with other sources of data, including optical satellite imagery and satellite-based AIS, to identify events and deduce responsibility. KSAT's system has been employed by customers all over the world, from national authorities to the offshore oil and gas industry, for self-monitoring, early spill detection, legal defence (in the case of false accusations), identification of polluters, regulatory compliance, and documentation of baseline conditions (KSAT 2016).

1.3.2.2 Gulf of Mexico Oil Spill

In 2010, the MODIS sensor on NASA's Terra/Aqua satellites provided a unique twice-daily perspective of the extent and movement of the oil slick originating from the Deepwater Horizon offshore oil platform (Fig. 1.4). Scientists and emergency response personnel used satellite imagery, in combination with ground observations and aerial photography, to plan and assess the progress of clean-up efforts. Satellite imagery was used to estimate the extent of ocean impacted by the disaster as 68,000 square miles (180,000 km^2). Such impact assessments are crucial inputs for legal proceedings and damage claims.

1.3.2.3 Red Tides

Red tides are algal blooms (phytoplankton) that occur naturally given certain combinations of environmental conditions. They are sometimes intensified by increased nutrient loading and temperature increases arising from human activities such as agricultural fertilization and coastal industrial activity. Harmful Algal Blooms (HABs) produce toxins that are dangerous to marine life and humans. Figure 1.5, captured by AVNIR-2 on the Japan Aerospace Exploration Agency's (JAXA) ALOS satellite, shows a red tide that occurred in Tokyo Bay from spring to summer of 2006. Satellite observations can be used to detect outbreaks before they become large in scale, facilitating management activities and identification of contributing factors.

Fig. 1.4 Deepwater Horizon oil spill captured by MODIS (NASA MODIS Rapid Response Team 2010)

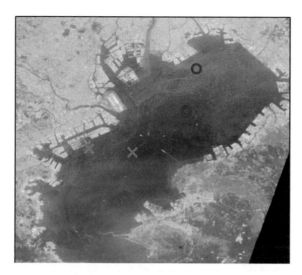

Fig. 1.5 JAXA ALOS (AVNIR-2) image showing a *red* tide that occurred in Tokyo Bay from spring to summer in 2006 (JAXA 2016a, b, c, d)

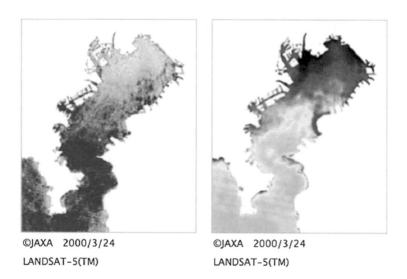

©JAXA 2000/3/24   ©JAXA 2000/3/24

LANDSAT-5(TM)   LANDSAT-5(TM)

Fig. 1.6 Turbidity distribution (*left*) and clarity distribution (*right*) (JAXA 2016a, b, c, d)

Figure 1.6 shows some other useful products derived from satellite imagery—turbidity distribution (*left*) and clarity distribution (*right*). This information can be used to plot red tide distributions and inform decision-makers who must assess water quality and safety.

1.3.3   Deforestation

In response to Brazilian Amazon deforestation, the Japan International Cooperation
Agency (JICA) established a technical cooperation project with the Brazilian
Institute for the Environment and Renewable Natural Resources (IBAMA) to
combat illegal deforestation. Using data from JAXA's ALOS and ALOS-2 SAR
instruments (e.g., Fig. 1.7), which have the ability to image in all weather conditions and regardless of the time of day, Brazilian authorities assess deforestation
and work with local law enforcement agencies to control illegal activities. Over the
2010–2012 period, IBAMA reported a 40% decrease in deforestation, due in part to
the collaboration with JICA and JAXA. These activities play a vital role in
achieving Brazil's Nationally Determined Contribution (NDC) to combating climate change, which aims to reduce the rate of Brazilian Amazon deforestation 80%
by 2020 (with respect to the 1996–2005 baseline of 19,500 km^2).

Figure 1.8 (from Landsat 4 TM in 1988) shows the drastic impact that politics
and social forces can have on the environment. The image shows heavy agricultural
development and thus forest clearing on the Mexican side of the border, which may
have been due to the relative political stability in Mexico at the time, whereas
Guatemala—in the midst of a civil war—experienced much less agricultural
development and forest clearing. This image spurred bilateral discussions between
the countries leading to increased conservation efforts, including the establishment
of the Maya Biosphere Reserve in 1990 (Simmon and Gray 2012).

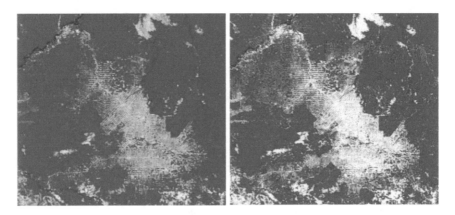

Fig. 1.7 Significant numbers of trees have been felled in Brazil, as evidenced by these two images
acquired by Japan's JERS-1 (*left*) and ALOS (*right*) satellites in 1995 and 2007 respectively
(Shimada 2016)

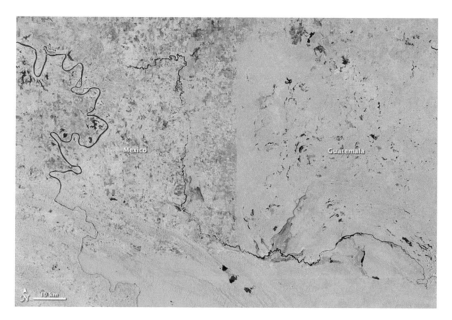

Fig. 1.8 Forest cover (*green*) at the border of Mexico and Guatemala (Simmon and Gray 2012)

1.4   A Taxonomy of Roles for Satellite Earth Observations

The examples highlighted in the preceding section make it clear that there is a place for satellite observations in efforts to solve a variety of environmental problems arising at the international level. However, they do not provide a more systematic account of the variety of roles that satellite observations can play in this realm. In this section, we tackle this challenge directly, presenting a simple taxonomy of the major roles that satellite observations can play in addressing environmental problems. This taxonomy offers an overview of the range and diversity of the contributions that satellite observations can make. It also provides a roadmap for those desiring to identify specific roles that can be subjected to more detailed analysis.

Table 1.1 provides a framework for thinking more systematically about roles for satellite observations. The table draws attention to environmental problems dealing with atmospheric, marine, and terrestrial systems. To some extent, these distinctions are artificial. Runoff of pesticides and fertilizers used in terrestrial farming operations, for example, is the principal source of marine dead zones. Tsunamis are marine phenomena, but they cause terrestrial environmental problems when the resultant tidal waves make landfall with destructive force. Nevertheless, it is helpful to begin with these distinctions in order to construct an overview of the roles of satellite Earth observations.

Table 1.1 Roles for satellite Earth observations in addressing environmental concerns

Function		Medium		
		Atmospheric	Marine	Terrestrial
Inform	Identify	Existence of the ozone hole	Occurrence of marine dead zones	Forest fires
	Monitor	Greenhouse gas concentration	Sea surface temperature	Pace of Amazon deforestation
	Assess	Recovery of the ozone layer	Rise of sea level Loss of sea ice	Loss of carbon stock
Assist		PM2.5 early warning	Provide early warning of tsunamis	Track pathways of tornadoes
Comply		Identify sources of greenhouse gas (CO_2 and CH_4) emissions	Track Illegal, Unreported, and Unregulated (IUU) fishing	Locate illegal loggers

Similarly, the table differentiates a number of distinct roles for satellite observations in coming to terms with environmental problems. The fundamental distinction separates roles that center on the supply of information, roles that involve the provision of emergency assistance, and roles that contribute to the achievement of compliance. Here, too, the boundaries are not watertight. Monitoring trends in concentrations of greenhouse gases in the Earth's atmosphere, for example, may prove helpful for those interested in the extent to which major countries are taking the steps needed to implement their NDCs under the terms of the 2015 Paris Agreement regarding climate change. Nevertheless, the categories included in the table do point to differences in purposes and in targets of observation that are important when it comes to the administration of Earth observation systems.

1.4.1 *Identify*

To begin with, satellite Earth observations can play critical roles in assembling the data needed to detect large-scale environmental problems, monitor shifts in the status or severity of these problems over time, and assess the success of concerted efforts to solve them. In some cases, the problems are largely unknown prior to observations from satellites. In the case of deforestation in the Amazon Basin, for instance, ground-based data were available to document local occurrences of destructive practices, but no one was able to grasp the extent of deforestation prior to the advent of remote-sensing images (National Research Council 1998). In other cases, satellite observations played a role in confirming the existence of a problem first identified by other means. The seasonal thinning of the stratospheric ozone layer, for example, was identified initially through ground-based observations made by British Antarctic Survey scientists located at Halley Bay, but satellite observations were able to verify these measurements and document the full scope and severity of the problem. In such cases, satellite observations can alter policy agendas by introducing and highlighting the significance of issues that did not exist previously as policy concerns, whether or not the problems had already come into existence in biophysical terms.[4]

1.4.2 Monitor

Once an environmental problem is identified, it becomes important to monitor the evolution of the problem over time. Satellite observations can play two distinct roles in this regard. One role involves monitoring trends in key variables without reference to policy interventions. The use of satellite observations to track shifts in seasonal maxima and minima in the extent of sea ice in the Arctic Basin or seasonal maxima and minima in the extent of dead zones in the Gulf of Mexico exemplifies this role. The other role centers on tracking progress toward fulfilling goals included in international treaties or agreements. This role is referred to as "systematic observations" in the ozone and climate change treaties (United Nations 1985a, b, 1992a, b). This is not a matter of measuring compliance on the part of individual actors. Rather, the critical concern is to contribute to determining whether the creation and implementation of an international regime is making a difference with regard to the status of the relevant problem. For example, are the so-called ozone holes over the polar regions becoming more or less severe over the course of time? Are glaciers retreating and, if so, is the rate of retreat increasing? Is the rate of desertification in sub-Saharan Africa or in northwestern China accelerating or slowing over time? In each case, the challenge is to explore links between the operation of a regime and observed trends in relevant biophysical phenomena.

---

[4]Litfin (1995) stated that the discovery of the Antarctic ozone hole (published in the journal Nature by Josef Farman of the British Antarctic Survey in its paper about severe ozone depletion in the Antarctic and a strong correlation between CFC concentrations and ozone losses) was "officially ignored" in the international negotiations for the Montreal Protocol. However, "the ozone hole, signalling a dangerously high probability of ecological disaster, precipitated a sense of crisis conductive to the precautionary discourse eventually sanctioned in Montreal", and "once the hole's existence was confirmed by NASA satellite data, the race was on to explain it." However, the hole was unexplained until aircraft based measurements could link the hole to CFCs, and thus, the fact that the hole remained unexplained until after the Montreal Protocol was signed, indicates that "scientific ignorance, rather than scientific knowledge, set the stage for international cooperation".

1.4.3 Assess

Going a step further, satellite observations can play an important role in evaluating the effectiveness of international agreements at solving the problems leading to their negotiation. The goal of the 2015 Paris Agreement is framed in terms of temperature increases at the Earth's surface. But to achieve this goal, it will be essential to limit increases in concentrations of greenhouse gases in the atmosphere. Satellite observations will play an essential role in assessing trends in these concentrations. Similarly, the program known as Reducing Emissions from Deforestation and Forest Degradation in Developing Countries (REDD+) focuses on initiatives to provide developing countries with incentives to reduce emissions from forested lands or to increase areas devoted to forests. Satellite observations will play an important role in assessing the performance of this program over time. The point here is not to document (non-)compliance with rules or obligations on the part of individual members of international regimes. Rather, the emphasis is on evaluating the performance of the regime as a whole over time, with the objective of determining whether there is a need to adjust or reformulate the regime going forward in the interest of solving the relevant problem.

1.4.4 Assist

A different type of role for satellite observations involves provision of assistance regarding matters like early warning and search and rescue. This role encompasses a range of more practical contributions in situations involving natural or anthropogenic disasters. The International Charter on Space and Major Disasters aims at providing a unified system of space data acquisition and delivery to those affected by natural or man-made disasters through Authorized Users. Each member space agency has committed resources to support the provisions of the Charter and is thus helping to mitigate the effects of disasters on human life and property by responding to requests from afflicted countries for imagery. One of the latest applications involved the flood that occurred in Sri Lanka on May 17, 2016. Satellite observations can play major roles in providing early warning regarding the occurrence of tidal waves produced by tsunamis and the tracks of hurricanes and tornadoes. A well-developed space-based system would have been able to provide significant early warning in the case of the December 26, 2004 Indian Ocean earthquake and tsunami, an event that is estimated to have killed over 300,000 people. Similar comments are in order regarding search and rescue. AISs using low Earth orbit satellites sometimes in combination with observation imagery are capable of tracking ships at sea in real time and guiding rescuers to the site of maritime disasters on an efficient basis. On an operational basis, meteorological services provided by geostationary and other satellites around the world provide a successful model of internationally coordinated and shared information services using satellites for weather forecasting and management of extreme weather, as well as for climate monitoring services.

1.4.5 Comply

Taking the next step, satellite observations become more sensitive because they focus on the extent to which individual subjects (both states and non-state actors) comply with requirements and prohibitions rather than on systemic concerns, such as the size of ozone holes or rates of desertification. AISs, for example, can identify the precise location of ships at sea and access relevant databases to determine whether a given ship is in possession of an up-to-date certificate or license to operate in the relevant area (e.g., a Polar Certificate issued in accordance with the International Maritime Organization's new rules for ships operating in polar waters). Recent advances are likely to allow satellites to make an accurate determination of whether a ship is engaged in illegal, unreported, and unregulated (IUU) fishing, even in cases where the operators of a ship have turned off their transponder in an effort to avoid detection. Remote sensing may provide the only way to document the occurrence and the extent of illegal logging in remote areas such as the islands of Borneo and Kalimantan. Several countries have started to make regular use of satellite data for their national inventories of greenhouse gas emissions (by observing forest areas as carbon sinks); the information will be used for self-reporting according to the obligations included in the Paris Agreement (United Nations 1992a, b, 1997, 2015). Innovations in the technology underlying satellite observations are occurring regularly. If the proposition that such observations constitute a legitimate compliance mechanism is widely accepted, innovations that are targeted to this role of satellite observations are likely to be forthcoming.

1.5 Coordination Mechanisms

There have been several multilateral initiatives to coordinate the various national Earth observation programs and policies. Among these are intergovernmental programs, including the Global World Weather Watch (WWW) (World Meteorological Organization 2016) and Earthwatch (United Nations Environment Programme 2016),[5] informal voluntary groups or partnerships such as the Committee on Earth Observation Satellites (CEOS), the Global Earth Observation

---

[5]The Global Resource Information Database (GRID) under the framework of Earthwatch integrates satellite remote sensing data and data collected by the Global Environment Monitoring System (GEMS).

System of Systems (GEOSS), and a number of other initiatives. The earliest efforts of multilateral coordination in remote sensing were with meteorological satellites. The WWW, the principal activity of the World Meteorological Organization (WMO), is a cooperative program for collecting, processing, and disseminating meteorological data from satellites and other sources, aiming to maximize the utilization of meteorological data from satellites. The Coordination Group for Meteorological Satellites (CGMS) meets annually to coordinate technical standards among satellite operators.

Intergovernmental agencies affiliated with the UN play a significant role in these initiatives for multilateral coordination of Earth observation and research for the protection of the environment. Among these are the World Climate Research Programme (WCRP), which studies physical aspects of climate change, and Future Earth, which operates under the auspices of the International Council for Science (ICSU). The Intergovernmental Oceanographic Commission (IOC), UN Environment Programme (UNEP), the UN Educational, Scientific, and Cultural Organization (UNESCO), and the WMO also help in planning these international research efforts. Funding agencies, such as the International Group of Funding Agencies for Global Change Research (IGFA) and what is now known as the Belmont Forum, also play an important role. To respond to the need for long-term climate monitoring, the Global Climate Observing System (GCOS) was established as a user-driven operational system capable of providing the comprehensive observations required for monitoring the climate system; detecting and attributing climate change; assessing impacts of, and supporting adaptation to, climate variability and change; application to national economic development; and research to improve understanding, modeling, and prediction of the climate system (GCOS 2016).[6] The Global Ocean Observing System (GOOS) is a permanent global system for observations, modeling, and analysis of marine and ocean variables to support operational ocean services worldwide (GOOS 2016). A central purpose of these research programs is to inform and influence national policies and international agreements on environmental management.

In the first instance, coordination is a matter of avoiding unnecessary duplication of efforts and ensuring that there are no gaps in coverage regarding key roles played by satellite observations. At the same time, it is important to agree on rules and procedures (e.g., format standardization, data principles, and sometimes joint program planning), allowing for healthy competition among Earth-observing programs to encourage the flow of innovations that can expand the capacity of space agencies to play constructive roles in dealing with a variety of environmental problems.

Coordination is also a matter of linking the contributions of a variety of Earth observation instruments, including ocean buoys, meteorological stations and balloons, seismic and Global Positioning System (GPS) stations, remote-sensing satellites, computerized forecasting models, and early warning systems (Group on Earth Observations 2008). For most of the roles identified in the preceding section,

---

[6]The GCOS is sponsored by: The WMO, UNESCO and its IOC, UNEP, and ICSU.

there is a need to integrate data coming from multiple sources to achieve the clearest and most detailed picture of the targets of Earth observations.

Many environmental agreements contain provisions for research and systematic observations, monitoring, or scientific research cooperation. The 1985 Vienna Convention for the Protection of the Ozone Layer expressly mentions satellite measurement (United Nations 1985a, b). The World Summit on Sustainable Development held in Johannesburg in 2002 adopted a Plan of Implementation, which includes several proposals on actions for satellite Earth observation, global mapping, and integrated global observations (United Nations 2002). At the same time, efforts to identify the adequacy of global observations for tracking climate change (GCOS 2003)[7] were carried out by GCOS and have been reported to the COP of the UNFCCC. The U.S. took the initiative to host the first ministerial Earth Observation Summit in Washington D.C. in 2003. In response to a call from the Group of Eight leading industrialized countries of June 2003 (G8 2003), the third Earth Observation Summit took place in Brussels in February 2005, initiating the formation of the GEOSS.

GEO is a voluntary partnership of governments and international organizations.[8] It provides a framework for these partners to develop new projects and coordinate their strategies and investments. GEO

> ...links existing and planned Earth observation systems and supports the development of new ones in cases of perceived gaps in the supply of environment-related information. It aims to construct a global public infrastructure for Earth observations consisting in a flexible and distributed network of systems and content providers. (International Institute for Sustainable Development 2015)

GEO has taken a leading role in coordinating efforts to build GEOSS. At its 12th Plenary Session, held in Mexico City in November 2015, GEO entered its second decade and adopted a new Strategic Plan covering the decade from 2016 to 2025 (for more details see Chap. 11).

1.6 The Governance of Earth Observation Systems

As satellite observations have taken on expanded roles in responding to the provision of emergency assistance, the operation of compliance mechanisms, and the assessment of regime performance, the need to move beyond simple procedures to coordinate the activities of producers of Earth observation data has come into focus. Consider the following development in this context. In November 2014, Skytruth, Oceana, and Google announced the launch of Global Fishing Watch, a public tool utilizing SpaceQuest AIS data and algorithms developed by Analyze Corp to

---

[7]Hereinafter: *GCOS Second Adequacy Report.*

[8]Some regard GEO as being technically an intergovernmental meeting but virtually a small-scale intergovernmental organization, see: Aoki (2006).

identify and display fishing activity worldwide.[9] Turning to hypothetical but entirely realistic scenarios, consider a case in which SpaceQuest, a private company, launches a satellite on a Russian rocket with the intention of deploying an AIS system to monitor compliance with the rules articulated in the IMO's Polar Code for ships operating in polar waters and to provide relevant data to governments in the countries where the ships are registered or where there are ports the ships are likely to enter. Or, to take another example, consider a partnership in which the European Space Agency (ESA) joins forces with Australia to track and ultimately apprehend a vessel registered in Panama that is thought to be harvesting Patagonian toothfish illegally in waters subject to the management regime established under the terms of the Convention on the Conservation of Antarctic Marine Living Resources (CCAMLR).

As these examples suggest, we are moving into a realm regarding the role of satellite observations in addressing environmental problems that takes us beyond the efforts of consortia of producers like GEO. We are now dealing with a complex mix of private companies, public agencies, and public/private partnerships. The private companies are interested in selling their services to a variety of users that may include academic, commercial, governmental, and non-governmental customers. This is certainly reasonable in principle, though it may require focused and careful efforts to iron out differences among participants regarding the proper division of labor between private sector actors (e.g., SpaceQuest or Google Earth) and public sector players like NASA, ESA, or JAXA and many other existing and emerging space agencies around the world.

Note also that the relevant partnerships may involve representatives of intergovernmental organizations responsible for administering international regimes, such as the IMO and the secretariat of CCAMLR. While the IMO does not have the capacity to operate Earth-observing satellites of its own, it does have the authority to make and interpret international rules pertaining to various aspects of the design, construction, and operation of commercial vessels. As the toothfish example suggests, there may be cases in which it is important to coordinate the efforts of providers of Earth observations not only with the activities of states (e.g., Australia) keen on improving compliance with international rules but also the responses of states (e.g., Panama) that may be less enthusiastic about the application of the rules to actors operating under their jurisdiction (e.g., vessels registered in Panama).

Another important observation regarding the roles of satellite observations in addressing large-scale environmental problems is that these roles are normally instrumental. As our taxonomy of roles indicates, we are concerned with matters like the depletion of the stratospheric ozone layer, the growth of marine dead zones, and the destruction of tropical forests. Satellite observations cannot solve these problems. But they can play a variety of supporting roles that are helpful to those responsible for coming to terms with these problems.

---

[9]http://globalfishingwatch.org/.

This makes it clear that there is a need for close cooperation between those seeking to implement or administer issue-specific regimes and those responsible for the operation of Earth-observing systems. It is possible, in principle, to launch satellites dedicated to the concerns of specific regimes and operated under the auspices of each regime. But arrangements of this sort are unlikely to become commonplace, for several reasons. Operating Earth-observing satellites requires material resources and trained personnel that are not available to most of those responsible for the administration of issue-specific regimes. There is no prospect that this will change in the foreseeable future. In addition, there are cases in which a satellite or a set of satellites can provide data responsive to the needs of two or more issue-specific regimes.

Under the circumstances, a key issue of governance will center on arrangements designed to ensure compatibility between the needs of the users of satellite observations and the activities of the providers of these data. Needed in this connection are what commentators on environmental governance call regime complexes (Oberthür and Stokke 2011). Such complexes are sets of institutional arrangements that deal with the same issues or with overlapping issues but that are not related to one another in a hierarchical manner. In such cases, it becomes important to work out, either formally or informally, a set of practices governing the interactions among the elements of the complexes.

There are examples of such arrangements that may offer insights to those concerned with the roles of Earth-observing systems. The Intergovernmental Panel on Climate Change (IPCC), for instance, is a body operating under the auspices of WMO and UNEP (Agrawala 1998). It provides scientific assessments that feed into the work of the UNFCCC, but it is not subject to the authority of the COP of the UNFCCC. The Intergovernmental Science-Policy Platform on Biodiversity and Ecosystem Services (IPBES) operates under the auspices of UNEP, the UN Development Programme (UNDP), the UN Food and Agriculture Organization (FAO), and UNESCO (Díaz 2015). It provides scientific input in response to requests from decision-makers, especially those responsible for the administration of international governance systems like the Convention on Biological Diversity.

The situation regarding Earth-observing systems is more complex (see Chap. 16). But it is easy to identify some of the key issues arising in the relevant regime complexes. Consider the following examples.

At the center of these arrangements lie relationships between the providers and users of Earth observation data. It is therefore essential to consider the terms of trade between upstream providers and downstream users. Should Earth observations be supported by public funds and made available to all (or all certified) users on an open access basis? What is the role of private providers in such settings? Can users make contracts with providers to meet the needs of users on an ongoing basis under terms agreed upon for varying lengths of time? What are the incentives for providers to develop technological innovations that can improve the services they are able to supply to users of satellite observations?

Are there issues regarding the protection of privacy or proprietary information that need to be addressed in the production and dissemination of satellite observations that can help to solve environmental problems? With current satellite technology, it is possible to obtain images of activities occurring on the Earth's surface with a very high resolution. Using Google Earth, for instance, it is possible to obtain detailed views of individual houses located on a particular street within a specific community. Additional technological advances will make it possible to reveal even more details regarding the activities of various actors. This is good news in some respects. For example, providers or disseminators of satellite images now say that they can identify fishing vessels even when they have their transponders turned off and determine whether they have nets in the water in violation of the rules governing specific areas or seasons.[10] But it may be feasible to place certain restrictions on the acquisition of satellite images in order to protect privacy, without diminishing the usefulness of the relevant data for the implementation of specific regimes. To take a concrete example, it is possible to determine whether a commercial vessel operating in polar waters has a certificate that is valid under the terms of the Polar Code without tracking the movements of the vessel continuously in real time and seeking to determine its cargo.

Should we be concerned about dangers relating to the misuse of Earth observation data for purposes that are unrelated to solving environmental problems? The same data, especially in the case of very high resolution data, can be used to address environmental problems and for military purposes. Today, not only do many countries have space programs, but very high resolution data are also openly available through the expanding global market. Many countries operating Earth observation satellites or that have private businesses operating satellites are aware of the risk of these data being used against their security interests. Therefore, in many cases, regulations are adopted to control the acquisition of very high-resolution satellite data. As experience in the realm of arms control makes clear, the danger that satellite observations will be misused for purposes that threaten national security is a real concern in all cases. Although this problem seems less serious when it comes to addressing large-scale environmental problems, it would be a mistake to assume that there is no reason to be alert to concerns of this sort.

Such issues concerning data rights, security, and privacy have been dealt with for decades through what is commonly called Earth observation data policy. As difficult as it is to tackle these problems effectively, the overall message of this section is that we need to consider a transition from a continuing concern with coordination to a new awareness of the need for governance in the use of satellite observations to help in solving environmental problems. The activities of groups like GEO and CEOS, which focus on coordination among producers of Earth observation data, will continue to be important; they may well become increasingly important. But it will be necessary to supplement these efforts with an appropriate set of rules and practices addressing issues that arise in interactions between

---

[10]See Howard (2015) for a report on the development of a collaboration involving SkyTruth, Oceana, and Google.

producers and users and, more specifically, in the contributions of satellite observations to the work of a range of issue-specific regimes addressing atmospheric, marine, and terrestrial problems. As a preliminary response to this need, efforts are taking place around the world to demonstrate the socioeconomic benefits of Earth observations. Space agencies and governments are faced with increasing pressure to plan space programs on evidence-based studies of anticipated societal benefits from investments in Earth observation systems. Such efforts should have in view the transition they want to achieve, which is to reach a world where satellite observations provide the data and information that meet the needs of policymakers through a set of rules and institutions that allow this to happen. The key to moving from technology driven R&D to societal benefits depends on whether or not this transition happens.

1.7 Architecture of the Book

The three substantive sections of this book contain a series of chapters that explore methodological issues relating to the development of policy-relevant satellite observations in a number of issue areas, the organization of agencies engaged in this work in both national and international settings, and the coordination and governance of Earth-observing systems to maximize their contributions to solving a range of environmental problems.

Part II on A Study on Methods for Assessing the Impact of Satellite Observations on Environmental Policy (Japan) describes the work of the Japanese Policy and Earth Observation Innovation Cycle (PEOIC) project. It includes a case study of the roles that satellite observations have played in supporting efforts to protect the stratospheric ozone layer. Part III turns to national and regional experiences in the U.S., Europe, and Asia. In addition to the experiences of NASA, ESA, and JAXA, it includes assessments of the UK Earth Observation Policy, the French Earth Observation Strategy, the Chinese Earth Observation Programme, and a specific case on greenhouse gas observation from space. Part IV deals with international initiatives and efforts to use satellite observations in coming to terms with issues of international significance. Specific topics include the experience and plans of the Organisation for Economic Co-operation and Development (OECD), Space Forum and GEOSS. Applications considered relate to the uses of satellite observations in dealing with matters of public health, the implementation of REDD +, and other global initiatives relating to the protection of forests and wetlands.

Part V on Prospects and Conclusions includes chapters that are more forward-looking regarding the roles of satellite observations in solving large-scale environmental problems. Chapter 16 focuses on future directions in institutional and organizational arrangements. It includes an analysis of whether there is a need to move toward the development of an arrangement that can be characterized as an "Earth observation regime complex" or an Earth observation component in various regime complexes dealing with large-scale atmospheric, marine, and terrestrial problems.

Chapter 17 is devoted to conclusions and recommendations. It addresses the principal conclusions emerging from a workshop of the PEOIC Advisory Board held in Tokyo, November 9–10, 2015, and seeks to draw on the cases discussed in the previous parts to extract lessons, and to articulate and refine a set of recommendations regarding ways to maximize the contributions of Earth observation systems both to policymaking in various issue areas and to addressing societal needs for environmental protection more generally.

Acknowledgements We thank Dr. Tatsuya Yokota and Dr. Akihiko Kuze for their technical suggestions on greenhouse gas observations.

References

Agrawala S (1998) Structural and process history of the intergovernmental panel on climate change. Clim Change 39(4):621–642

Aoki S (2006) Nihon no uchu senryaku (Japanese Space Strategy). Tokyo, Japan

Colwell RN (1983) Manual of remote sensing. Falls Church, Virginia

Díaz S (2015) The IPBES conceptual framework — connecting nature and people. Curr Opin Environ Sustain 14:1–16

G8 (2003) The Group of Eight Industrialized Nations, science and technology for sustainable development – a G8 action plan. http://www.g8.fr/evian/english/navigation/2003_g8_summit/summit_documents/science_and_technology_for_sustainable_development_-_a_g8_action_plan.html. Accessed 2 Dec 2008

GCOS (2003) The second report on the adequacy of the global observing systems for climate in support of the UNFCCC. http://www.wmo.int/pages/prog/gcos/Publications/gcos-82_2AR.pdf. Accessed 6 Dec 2016

GCOS (2016) GCOS: background. http://www.wmo.int/pages/prog/gcos/index.php?name=Background. Accessed 30 May 2016

GOOS (2016) What Is GOOS? http://www.ioc-goos.org/index.php?option=com_content&view=article&id=12&Itemid=26&lang=en. Accessed 30 May 2016

Group on Earth Observations (2008) GEO information kit – information sheet 2: global Earth observation system of systems. https://www.earthobservations.org/documents/geo_information_kit.pdf. Accessed 6 Dec 2016

Howard BC (2015) Tiny team uses satellites to bust illegal fishing worldwide. http://news.nationalgeographic.com/2015/06/150615-skytruth-pirate-fishing-illegal-big-data-ocean-conservation/. Accessed 6 Dec 2016

International Institute for Sustainable Development (2015) A Summary report of the 12th plenary session of the group on Earth observations (GEO-XII) and GEO 2015 Mexico City ministerial summit. http://www.iisd.ca/download/pdf/sd/enbplus220num5e.pdf. Accessed 6 Dec 2016

JAXA (2016a) Marine Pollution | Monitoring the Global Environment | JAXA Earth Observation Research Center (EORC). http://www.eorc.jaxa.jp/en/observation/global/marine_pollution.html. Accessed 15 Apr 2016

JAXA (2016b) Mechanism of remote sensing. http://www.eorc.jaxa.jp/en/hatoyama/experience/rm_kiso/mecha_howto_e.html. Accessed 1 Aug 2016

JAXA (2016c) Red tide around Japan from spring to early summer in 2006 captured by AVNIR-2. http://www.eorc.jaxa.jp/ALOS/en/img_up/ex_akashio2006/akashio_01.htm. Accessed 15 Apr 2016

JAXA (2016d) Whole-atmospheric monthly CO_2 concentration tops 400 ppm – preliminary GOSAT monitoring results. http://global.jaxa.jp/press/2016/05/20160520_ibuki.html. Accessed 28 Jul 2016

JAXA/NIES/MOE (2016) Recent global CO_2. http://www.gosat.nies.go.jp/en/recent-global-co2.html. Accessed 30 May 2016

KSAT (2016) Oil slick detection service. http://www.ksat.no/en/services%20ksat/oil%20slick%20detection%20service%20-%20page/. Accessed 30 May 2016

Litfin KT (1995) Framing science: precautionary discourse and the ozone treaties. Millenuium J Int Stud 24(2)

Mather PM (1999) Computer processing of remotely-sensed images: an introduction, 2nd edn. Wiley, Chichester

NASA MODIS Rapid Response Team (2010) Gulf of Mexico oil slick images: frequently asked questions: feature articles. http://earthobservatory.nasa.gov/Features/OilSlick/. Accessed 15 Apr 2016

NASA/JPL-Caltech (2015) Excitement grows as NASA carbon sleuth begins year two—Orbiting Carbon Observatory. http://oco.jpl.nasa.gov/newsoco/index.cfm?fuseaction=ShowNews&newsID=162. Accessed 15 Apr 2016

National Research Council (1998) People and pixels: linking remote sensing and social science. Natonal Academy Press, Washington, DC

OCO-2 Science Team/Gunson M, Eldering A (2015) OCO-2 Level 2 bias-corrected XCO_2 and other select fields from the full-physics retrieval aggregated as daily files, retrospective processing V7r, version 7r. http://disc.sci.gsfc.nasa.gov/datacollection/OCO2_L2_Lite_FP_7r.html. Accessed 24 Aug 2016

Shimada M (2016) Quick deforestation monitoring system using ALOS-2/PALSAR-2. http://ceos.org/document_management/Ad_Hoc_Teams/SDCG_for_GFOI/Meetings/SDCG-9/19_SDCG9_GFOI%E3%83%97%E3%83%AC%E3%82%BC%E3%83%B3%E8%B3%87%E6%96%9920160225-shimda-3.pdf. Accessed 15 Apr 2016

Short NM (2008) Historical and technical perspectives of remote sensing. http://www.fas.org/irp/imint/docs/rst/Intro/Part2_1.html. Accessed 7 Dec 2008

Simmon R, Gray E (2012) Landsat image gallery – border between Mexico and Guatemala. http://landsat.visibleearth.nasa.gov/view.php?id=78600. Accessed 15 Apr 2016

Oberthür S, Stokke OS (2011) Managing institutional complexity: regime interplay and global environmental change. MIT Press, Cambridge

The International Charter on Space and Major Disasters (2016) International Disasters Charter. https://www.disasterscharter.org/. Accessed 23 May 2016

United Nations (1985a) Vienna convention for the protection of the ozone layer, Art. 2.2 (a) and 3. United Nations Treaty Series, Vienna, Austria

United Nations (1985b) Vienna convention for the protection of the ozone layer, Art. 3, Annex I. United Nations Treaty Series, Vienna, Austria

United Nations (1992a) United nations framework convention on climate change, Art. 4.1 (a). United Nations Treaty Series, New York City, New York

United Nations (1992b) United nations framework convention on climate change, Art. 5. United Nations Treaty Series, New York City, New York

United Nations (1997) Kyoto Protocol to the united nations framework convention on climate change, Art. 3.3. United Nations Treaty Series, Kyoto, Japan

United Nations (2002) Plan of Implementation of the World Summit on sustainable development, resolution 2. United Nations, pp 7–72

United Nations (2015) Paris Agreement under the united nations framework convention on climate change, Art. 13.7 (a). United Nations Treaty Series, Paris, France

United Nations Environment Programme (2016) United nations system-wide earthwatch. http://earthwatch.unep.net/. Accessed 9 Nov 2008

World Meteorological Organization (2016) World weather watch. http://www.wmo.ch/web/www/www.html. Accessed 9 Nov 2008

Young O, King L, Schroeder H (2008) Institutions and environmental change. MIT Press, Cambridge, Massachusetts

Author Biographies

Oran R. Young is a renowned Arctic expert and a world leader in the fields of international governance and environmental institutions. His scientific work encompasses both basic research focusing on collective choice and social institutions, and applied research dealing with issues pertaining to international environmental governance and the Arctic as an international region. Professor Young served for 6 years as vice-president of the International Arctic Science Committee and was the founding chair of the Committee on the Human Dimensions of Global Change within the National Academy of Sciences in the U.S.A. He has chaired the Scientific Committee of the International Human Dimensions Programme on Global Environmental Change and the Steering Committee of the Arctic Governance Project.

Masami Onoda is currently the U.S. and multilateral relations interface at the International Relations and Research Department of the Japan Aerospace Exploration Agency (JAXA). As an academic, she is fellow of the Institute of Global Environmental Strategies (IGES) and she is also engaged in the private sector as an advisor to the Singapore-based space debris start-up Astroscale Pte. Ltd. since its foundation in 2013. From 2009 to 2012, Dr. Onoda was a scientific and technical officer at the intergovernmental Group on Earth Observations (GEO) Secretariat in Geneva, Switzerland. From 2003 to 2008, while pursuing her graduate studies, she was invited to the JAXA Kansai Satellite Office in Higashiosaka as a space technology coordinator to support technology transfer to SMEs for the small satellite project SOHLA-1. From 1999 to 2003, she worked in the field of Earth observations at JAXA (then NASDA), serving on the Secretariat of the Committee on Earth Observation Satellites (CEOS). In 1999, she was seconded to the UN Office for Outer Space Affairs (UNOOSA) for the organization of the UNISPACE III conference. She holds a Ph.D. in global environmental studies (2009) and a master's degree in environmental management (2005), both from the Kyoto University Graduate School of Global Environmental Studies. Her undergraduate degree is in international relations from The University of Tokyo.

Part II
A Study on Methods for Assessing the Impact of Satellite Observations on Environmental Policy (Japan)

Chapter 2
Policy and Earth Observation Innovation Cycle (PEOIC) Project (Japan)

Yasuko Kasai, Setsuko Aoki, Akiko Aizawa, Akiko Okamatsu, Tomohiro Sato, Masami Onoda and Brian Alan Johnson

2.1 Assessment Framework of the Policy and Earth Observation Innovation Cycle (PEOIC)

2.1.1 The Policy and Earth Observation Innovation Cycle

Earth observation has the potential to make significant contributions to policy and society, as we discussed in Chap. 1 of this book. In this chapter, we focus on the function of "Inform" and the medium "Atmospheric" as shown in Table 1.1 in Chap. 1. This role generally corresponds to global scale long-term monitoring by low Earth orbit (LEO) satellites. Through such observations, influential scientific findings from satellite data have triggered international debates and helped set the agenda for addressing many environmental problems, including stratospheric ozone depletion and global warming. On the other hand, the policy priorities of national

Y. Kasai (✉)
National Institute of Information and Communications Technology (NICT),
4-2-1, Nukui-Kitamachi, Koganei, Tokyo 184-8795, Japan
e-mail: ykasai@nict.go.jp

S. Aoki
Keio University Law School, 2-15-45 Mita, Minato-Ku, Tokyo 108-8345, Japan
e-mail: aosets@sfc.keio.ac.jp

A. Aizawa
National Institute of Informatics (NII), 2-1-2 Hitotsubashi, Chiyoda-Ku,
Tokyo 101-8430, Japan
e-mail: aizawa@nii.ac.jp

A. Okamatsu
Faculty of Sustainability Studies, Hosei University, 2-17-1 Fujimi, Chiyoda,
Tokyo 102-8160, Japan
e-mail: okamatsu@hosei.ac.jp

© The Author(s) 2017
M. Onoda and O.R. Young (eds.), *Satellite Earth Observations and Their Impact on Society and Policy*, DOI 10.1007/978-981-10-3713-9_2

governments are often the basis of their budgetary allocations to governmental space missions, with the goal being that satellite missions will lead to scientific findings that better inform decision-makers. We know from experience that such cases happen. Our purpose here is to show quantitative evidence of the relationship between observation data and policy.

In Japan, there is a growing demand to set mission goals for publicly funded satellite missions based on societal or policy outcomes. However, at present there are very few studies assessing the actual impacts of satellite Earth observations on policy. The Japanese government typically decides on the funding for satellite Earth observation missions through a one-way planning approach, where the scientific results/advancements from a mission do not fully contribute to the next step of the innovation process. The authors of this chapter think that this lack of analysis—either retrospective or prospective—of the impact of Earth observations on environmental policy is one reason why Japan's Earth observation programs lack consistent and continuous planning.

The goal of this research therefore is to develop methods for quantitative and objective assessment of the impact of satellite Earth observations on environmental policy. For this purpose, the research project members have proposed the concept of a "Policy and Earth Observation Innovation Cycle" (PEOIC). Satellite data are used to produce information and intelligence leading to strategy and the formation of policy. Ultimately, the outcome effect of the policy is monitored and this information feeds into the next cycle of scientific and technological innovation. The immediate goal of the project is to propose methodology options for Japan to perform such an assessment for better system planning in the future. The long-term goal beyond this project is to build a society where there is such a working cycle, which could be a key to developing the road map to Future Earth (2016). Figure 2.1 illustrates the PEOIC concept.

T. Sato
Big Data Analytics Laboratory, Big Data Integration Research Center, National Institute of Information and Communications Technology (NICT), 4-2-1, Nukui-Kitamachi, Koganei, Tokyo 184-8795, Japan
e-mail: tosato@nict.go.jp

M. Onoda
International Relations and Research Department, Japan Aerospace Exploration Agency, Ochanomizu Sola City, 4-6, Kanda Surugadai, Chiyoda-Ku, Tokyo 101-8008, Japan
e-mail: onoda.masami@jaxa.jp

B.A. Johnson
Natural Resources and Ecosystem Services, Institute for Global Environmental Strategies, 2108-11, Kamiyamaguchi, Hayama, Kanagawa 240-0115, Japan
e-mail: johnson@iges.or.jp

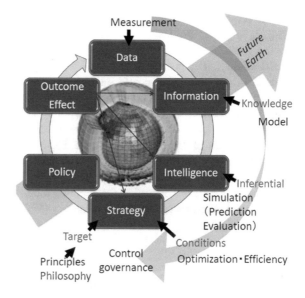

Fig. 2.1 Proposed "Policy and Earth Observation Innovation Cycle"

2.1.2 Project Outline

There are a number of studies that address the contributions or benefits of satellite remote sensing technologies to society and the relationship with policy (Booz and Company 2011; PricewaterhouseCoopers LLP 2006; Williamson et al. 2002; Macauley 2009; and others). These studies range from public reports to policy or economic analysis. In a broader context, the relationship between science and policy has been an important area of the study on state behaviors, in governance, or in the constructivist approaches in international relations.

We know that publicly funded satellite missions are commonly designed to meet the policy requirements that affect resource allocation. On the other hand, we also know that environmental policy agendas are set by scientific findings. Our investigation on the impact of satellite Earth observations on environmental policy has led us ultimately to the question of the role of science and policy and the interactions between the two, and whether evidence of such interactions could be identified in a quantitative manner. However, the lack of an approach to integrate policy, science, and technology seems to lead to a discrepancy between the two domains, causing inefficiencies in the system. If there is a way to identify and possibly quantify the data and information flow from science to policy, it should be possible to identify the existing gap and analyze the reason for its existence, leading to possible solutions to reconcile this discrepancy.

Based on these hypotheses, the project attempts to develop a tracking technique to link the currently disconnected fields of satellite Earth observation and environmental policy. The project further attempts to bridge the gap between policy and the development of science and technology research.

Our project started by developing a methodology for such an assessment by bringing together groups of specialists in different fields, including space and environmental law and policy, satellite engineering, and informatics. First, different phases of the policy process where satellite observations could have had an impact were identified through a document review; then text mining was performed to quantitatively assess if and how Earth observations had an impact on policy. The project comprises three components: the policy component, led by Prof. Setsuko Aoki of Keio University, the assessment methodology component, led by Dr. Masami Onoda of the Japan Aerospace Exploration Agency (JAXA), and the statistical analysis component led by Prof. Yasuko Kasai of the National Institute of Information and Communications Technology (NICT). The overall project leader is Prof. Kasai of NICT. The project was granted funding for three years (2013–2016) by the Japan Science and Technology Agency (JST) Research Institute for Science and Technology for Society (RISTEX).

2.1.3 Methodology and Results

First, our project conducted a survey to review the relevance of published literature related to each component of the research project, with the objective of developing a method to clarify the extent of the contributions of science, technology, and satellite data to policy through the assessment of literature and documents. The survey included a broad array of studies, from legal and institutional methods to economic methods (cost effectiveness analysis), and other literature on space and remote sensing policy and technology in general. The survey further focused on studies done in the U.S. and Europe that relate to satellite Earth observation and its impact on society in broad terms. The main findings were that: U.S. governmental Earth observation missions are mainly designed based on a survey of scientific requirements [i.e., the decadal survey for Earth science (NRC 2007)] in coordination with budget priorities; the European Global Monitoring for Environment and Security (GMES) program, now called Copernicus, was established based mainly on (operational) end user requirements and a prospective cost–benefit analysis of the program; very few policy documents or literature include specific references to satellite missions, instruments, or other details; existing studies focus on the socioeconomic benefits more than how satellite Earth observation impacts policy; the only closest methodology that used document analysis found during the survey was by Macauley (2009), which comprised an analysis on the use of Landsat data through searches of peer-reviewed scientific papers.

This survey demonstrated the limitations of assessing the impact of satellite data on policy by using data from document analysis, in that there were only a limited number of sample documents that could be used for the analysis. We also found that even if it is possible to analyze the relationship between satellite data and science, the relationship between science and policy is not a straightforward issue and is very difficult to quantify because of the many complicated factors involved.

In parallel, the project analyzed the case of satellite observations and the ozone regime, namely the Convention for the Protection of the Ozone Layer (United Nations 1985) (hereinafter the 'Vienna Convention') and the Protocol on Substances that Deplete the Ozone Layer (United Nations 1987) (hereinafter the 'Montreal Protocol'). The case was chosen because it is a successful example of an environmental agenda where satellite observations are said to have played a major role and was viewed as an ideal case study for retrospective quantitative analysis.

For this effort, Keio University's group researched the impact of satellite data on the amendment processes of the Montreal Protocol, using documents related to the Meeting of the Parties (MOP) and how satellite missions and instruments are mentioned in those documents. A team from the National Institute for Informatics (NII) worked with a team from JAXA and the Institute for Global Environmental Strategies (IGES) to develop a "dictionary" using the database of Earth observation missions developed by the Committee on Earth Observation Satellites (CEOS). The dictionary was then used to analyze how satellite data are being used in scientific papers and the public media (newspapers).

The NII research group worked with the NICT group to develop a document-sharing platform for analyzing the relevance between datasets from policy and from satellite observations. All groups worked together to populate the database for the platform. Using this platform, the case study on ozone gathered a number of policy-related documents, namely those documents from the Conference of Parties to the Vienna Convention and the Meeting of the Parties of the Montreal Protocol, and scientific resources such as the ozone assessments by the World Meteorological Organization (WMO) and the UN Environment Programme (UNEP). The documents were digitized and made available for electronic searching and mining. The details of this work are described in the following sections. This methodology is thought be adaptable to other issues, such as climate change or Arctic ocean navigation, in order to identify the role(s) of satellite data in those policy areas.

2.1.4   International Advisory Board and Future Prospects

As the project developed, the members decided that the idea of environmental governance could be a meaningful addition to existing research on Earth observation benefits. Since cost–benefit analysis has been one of the most commonly used methodologies for assessing the benefits of Earth observations in several countries/regions, it was thought that if we couple this with an institutional approach involving the role, rights, and duties of stakeholders, we might be able to establish a working cycle of policy and science.

In order for Earth observations and prediction systems to be used comprehensively for decision-making, a process must be established involving all stakeholders and end users. If we regard Earth observations as a tool to achieve the role of monitoring within the international governance system for the environment, appropriate institutions must be developed to achieve optimum use of Earth observations for environmental monitoring. To pursue this path, the project established an International Advisory Board chaired by Prof. Oran R. Young and composed of experts in socioeconomic studies and Earth and remote sensing sciences from around the world. A workshop was held in November 2015 in Tokyo, inviting the Advisory Board members to speak to the public as well as to participate in a workshop on the experiences of their different countries, regions, or organizations and how Japan could pursue its studies.

Taking into account such developments, the PEOIC project in its last year (2015–2016) started an effort to compile and categorize the present work of Earth observations benefits assessments around the world, and to develop criteria and recommendations for realizing an "Innovation Cycle." The project has asked Advisory Board members to submit contributions based on their own experience, using the template made for this study, and advise on how recommendations could be developed. Further, Profs. Oran R. Young and Olav Schram Stokke added some perspectives on a future "Earth Observation Regime Complex."

The following sections describe in detail the research conducted by each team. Section 2.2 addresses the work on the policy component led by Keio University to analyze the ozone regime and the use and impact of satellite Earth observations on the policy for the protection of the stratospheric ozone layer, extending this to the climate change regime. Section 2.3 by NICT and NII describes how data mining techniques were adopted to track the flow of information from satellite Earth observations to policy. Finally, Sect. 2.4 provides some conclusions for the project.

2.2 Protection of the Ozone Layer and Climate Change

2.2.1 Purpose and Methodology of the Case Study

In 1938, chlorofluorocarbons (CFCs) were discovered to be ideal chemicals for use as refrigerants, propellants and solvents, leading to their widespread use in many applications. However, Molina and Rowland (1974) published findings suggesting that CFCs contribute to ozone depletion in the upper atmosphere, which would result in severe consequences for humans and the Earth's ecosystems if left unchecked. The issue became a controversial topic of intense discussion in the scientific community and the findings were contested by the chemical industry, which denied the relationship between ozone depletion and CFCs. For example, Du Pont's congressional testimony said "The chlorine-ozone hypothesis is at this time purely speculative with no concrete evidence… to support it" and "If creditable scientific data… show that any chlorofluorocarbons cannot be used without a threat to health, Du Pont will stop production of these compounds."

Later it was also found that increased ozone in the troposphere is the third largest cause of global warming, only eclipsed by CO_2 and CH_4. Around the end of the twentieth century, it was also proven that global warming directly increases ozone transport from the stratosphere to troposphere, further increasing total ozone concentration, especially in mid-latitudes (WMO 1988a, b; IPCC 2001). A series of scientific findings suggested that it would not be enough nor efficient just to stop the ozone depletion resulting from human activities, but that an international regime to protect and preserve the appropriate climate status must be carefully constructed and maintained through global cooperation.

Such environmental and climate changes had been studied and supported by model simulations based on laboratory experiments, ground-based observations (especially those of Dobson and Brewer using spectrophotometers), and aircraft and balloon observations. However, satellite data constituted the only direct evidence of ozone layer destruction in the stratosphere as well as the increased transport and circulation of ozone between the stratosphere and upper parts of the troposphere (WMO 1988a, b). Ozone depletion in the stratosphere occurs by recombination with atomic oxygen ($O + O_3 \rightarrow 2O_2$), catalyzed by many atmospheric species such as OH, HO_2, NO and NO_2, or Cl and ClO. Satellite data provide convincing evidence that the international community can use to inform the necessary joint measures required to combat the situation. Data in the form of images are particularly impactful and tend to garner more attention and raise awareness of the international society.

This is evidenced in the confirmation of Molina and Rowland's 1974 hypothesis, which suggested that CFCs were destroying the stratospheric ozone layer. The Total Ozone Mapping Spectrometer (TOMS) on Nimbus-7 and the Solar Backscatter Ultraviolet (SBUV) instrument on NOAA-9 confirmed this hypothesis when they found striking evidence of an Antarctic "ozone hole" phenomenon in 1986 (NASA Ozone Watch 2013).

In 1977, the code of conduct for the ozone layer was adopted and the Coordinating Committee on the Ozone Layer (CCOL) was established. In the late 1970s, the U.S. began to ban the use of CFCs, but European countries opposed similar bans due to the cost of alternatives and a claimed lack of evidence. Many developing countries also opposed such measures, fearing repercussions to their industrial development.

Although industry strongly denied the link between CFCs and ozone depletion, the issue came to be discussed more frequently and Canada, Sweden, and Norway followed the lead of the U.S., enacting national legislations to ban the use of CFCs. In 1980, UNEP recommended that States should freeze or reduce the production and use of CFCs.

Finally, in 1985, the *Vienna Convention* was adopted, acting as a framework for international efforts to protect the ozone layer. However, the Vienna Convention does not include any legally binding reduction goals for CFCs. These targets are instead laid out in the accompanying *Montreal Protocol* adopted in 1987.

Today, the Montreal Protocol is arguably cited as: "the single most successful international agreement to date," as stated by Mr. Kofi Anan, the former Secretary General of the United Nations. This mechanism successfully eliminated the use of human-made ozone depleting substances (ODS) by January 1, 2010. It has been proven that the stratospheric ozone layer is now recovering (United Nations Environment Programme 2010).

While it is difficult to isolate the contribution of satellites, as scientific facts are identified using multiple approaches, this chapter studies whether satellite data had an influence on the decisions made during the implementation of the Montreal Protocol. For this purpose, Sects. 2.2.2 and 2.2.3 present the contents and developing obligations provided for in the Montreal Protocol and its later amendments. Section 2.2.4 explores whether a series of satellite instruments provided substantially important information on which a series of amendments to the Montreal Protocol were made. If this is answered in the affirmative, it would strengthen the basis for accelerating the development of Earth observation satellites through increased international cooperation. Conversely, if the answer is in the negative, it may suggest that there is room to reconsider the role of satellite Earth observations in support of climate policy.

2.2.2 Major Obligations of the Vienna Convention/Montreal Protocol System as an International Regime to Protect the Ozone Layer

The Vienna Convention defines the general obligation of the Parties to take appropriate measures to protect human health and the environment against adverse effects resulting from human activities that have been modifying the ozone layer (United Nations 1985). Parties to the Convention agreed to cooperate by means of systematic observations, research and information exchange, as well as assessment of the effect of human activities on the modification of the ozone layer. The main scientific issues recognized in the Vienna Convention were:

i. the modification of the ozone layer would result in a change in the amount of solar ultraviolet radiation having biological effects (UV-B) that reaches the Earth's surface and the potential consequences for human health, for organisms, ecosystems, and materials useful to mankind; and,

ii. the modification of the vertical distribution of ozone could change the temperature structure of the atmosphere and the potential consequences for weather and climate.

It was, therefore, specified in Annex I of the Vienna Convention that research into the physics and chemistry of the atmosphere would be conducted by means of not only laboratory studies and field measurements but also "instrument development, including satellite and non-satellite sensors for atmospheric trace constituents, solar flux and meteorological parameters." Likewise, it was planned that a Global Ozone Observing System of tropospheric and stratospheric concentrations of source gases for the HOx, NOx, ClOx, and carbon families would be constructed as an integration of satellite and ground-based systems. The following gases were specified in Annex I of the Vienna Convention as those to be monitored by the above-mentioned Global Ozone Observing Systems: carbon monoxide (CO), carbon dioxide (CO_2), methane (CH_4), non-methane hydrocarbon species from carbon substances; nitrous oxide (N_2O) and nitrogen oxides (NO_x) from nitrogen substances; halogenated alkanes including CCl_4, $CFCl_3$ (CFC-11), CF_2Cl_2 (CFC12), $C_2F_3Cl_3$ (CFC113), $C_2F_4Cl_2$ (CFC114), CH_3Cl, CHF_2Cl (CFC22), CH_3CCl_3, and $CHFCl_2$ (CFC21); CF_3Br (sources of B_rO_x) and hydrogen (H_2) in bromine substances; and hydrogen (H_2) and water (H_2O) in hydrogen substances. Note that the Vienna Convention only determined the kinds of gases to be monitored, not those to be reduced and ultimately eliminated for the consumption for each member state.

The Montreal Protocol explicitly designates the ODS that will be eliminated from production (United Nations 1987). This Protocol also provides for a grace period for developing countries due to the small amounts of ODS they emitted and their low capacity for developing alternatives to ODS. Annex A of the 1987 Montreal Protocol provides for the following substances: Group I chemicals ($CFCl_3$ (CFC-11), CF_2Cl_2 (CFC-12), $C_2F_3Cl_3$ (CFC-113), $C_2F_4Cl_2$ (CFC-114), C_2F_5Cl (CFC-115)), to be eliminated after January 1996; and, Group 2 chemicals (CF_2BrCl (Halon 1211), CF_3Br (Halon-1301) and $C_2F_4Br_2$ (Halon 2402)) to stop after 1 January 1994.

The Conference of the Parties (COP) held pursuant to the Vienna Convention may adopt additional protocols and any Party may also propose amendments to the Vienna Convention and to future protocols, taking "due account, inter alia, of relevant scientific and technical considerations." The mechanisms to counterattack the abundance of ODS in the atmosphere by the Vienna Convention/Montreal Protocol include a dynamic process to upgrade and modify the Parties' obligations in accordance with new scientific findings. In fact, as portrayed in the next section, the Montreal Protocol has been amended several times in response to the assessment of scientific findings. In addition, further funding mechanisms were introduced to assist developing countries in the process. This chapter will not address the capacity-building aspects of the ODS reduction strategy, but will focus on the relationship between the scientific findings, especially those obtained by satellite monitoring, and the change of obligations imposed on Parties to the Montreal Agreement (197 Parties as of April 2016). Considering the number of United Nations Member States (193 as of April 2016), it can safely be said that the Vienna Convention/Montreal Protocol regime represents one of the most established international regimes today.

2.2.3 Dynamic Obligations of ODS Elimination

2.2.3.1 Amendments and Adjustments of the Montreal Protocol

New scientific findings have led to changes in the obligations of the Montreal Protocol. Since entering into force in 1989, the Montreal Protocol has been amended ten times through two kinds of changes: narrow "amendments" and methodological "adjustments". The former means adding a new ODS to the list of substances to be eliminated. As this makes the Montreal Protocol more stringent than the original version, financial mechanisms have also been created to alleviate the burden on developing countries. The MOP to the Montreal Protocol adopted amendments to the Protocol four times. They are generally known by the names of the city where the annual MOP was held: the London Amendment (adopted in 1990, entered into force in 1992), the Copenhagen Amendment (1992, 1994), the Montreal Amendment (1997, 1999), and the Beijing Amendment (1999, 2002). As the traditional international technique is applied for the amendment, amended versions of the Montreal Protocol are applicable only to those States that have ratified them. As of April 2016, all 197 member States of the Montreal Protocol have ratified all four amended versions.

"Adjustment" refers to the acceleration of the reduction and elimination schedule in the production, consumption, and transfer of controlled substances that have already been designated in the original Montreal Protocol and its amended versions. Adjustments are possible if a two-thirds majority in the MOP is achieved, unless consensus is reached without a vote, and resulting adjustments are automatically applicable to all Parties. Adjustment becomes effective 6 months from the date of the notification by the Depositary. The Montreal Protocol has been adjusted six times as of April 2016: London (1990, 1991), Copenhagen (1992, 1993), Vienna (1995, 1996), Montreal (1997, 1998), Beijing (1999, 2000), and Montreal (2007, 2008).

2.2.3.2 ODS Added and Accelerated Through Amendments and Adjustments

London Amendment and Adjustment: The first Amendment to the Montreal Protocol was agreed upon by the second MOP held in London. This amendment entered into force 2 years later, in 1992. This amendment required a halt to the consumption or production of chemicals including a variety of CFCs, including carbon tetrachloride (CCl_4) and trichloroethane (methyl chloroform) after January 1, 1996 (United Nations Environment Programme Ozone Secretariat 1990). More specifically, the following ODS were added: F_3Cl (CFC-13), C_2FCl_5 (CFC-111), $C_2F_2Cl_4$ (CFC-112), C_3FCl_7 (CFC-211), $C_3F_2Cl_6$ (CFC-212), $C_3F_3Cl_5$ (CFC-213), $C_3F_4Cl_4$ (CFC-214), $C_3F_5Cl_3$ (CFC-215), $C_3F_6Cl_2$ (CFC-216), C_3F_7Cl (CFC-217), CCl_4, and $C_2H_3Cl_3$ (1,1,1-trichloroethane). The second MOP also adopted an accelerated schedule for the implementation of the CFCs (CFC-11, -12, -113, -114, -115) and Halons (Halon-1211, -1301, -2402), thus replacing Art. 2, paras. 1 and 2 with Art. 2A (CFCs) and Art. 2B (Halons) of the Montreal Protocol.

Copenhagen Amendment and Adjustment: The fourth MOP held in Copenhagen in 1992 added the following controlled substances: Hydrobromofluorocarbons (HBFCs) and CH_3Br (methyl bromide) (United Nations

Environment Programme Ozone Secretariat 1992). The same MOP succeeded in a further adjustment that accelerated the schedule of controlling the following substances: CFCs, Halons, other fully halogenated CFCs, CCCl$_4$ and 1,1,1-Trichloroethane (Methyl Chloroform).

Vienna Adjustment: While no amendment was accomplished in the seventh MOP held in Vienna in 1995, another adjustment was successfully decided, which became effective in 1996. The Vienna Adjustment accelerated the controlling schedule of HCFC and methyl bromide (United Nations Environment Programme Ozone Secretariat 1996a, b). Implementing the reduction ahead of schedule with respect to the following substances was also agreed upon vis-à-vis developing countries: CFC, Halon, Carbon Tetrachloride and 1,1,1-Trichloroethane (Methyl Chloroform) that had been previously only been the obligation of developed countries (United Nations Environment Programme Ozone Secretariat 1996a, b).

Montreal Amendment and Adjustment: The ninth MOP held in Montreal in 1997 did not add new substances to be eliminated; instead, the Amendment banned the trade of methyl bromide between the Parties and non-Parties, as well as providing further trade restrictions on new, used, recycled, and reclaimed controlled substances (United Nations Environment Programme Ozone Secretariat 1997a, b). The Adjustment accelerated the schedule of controlling methyl bromide, etc. (United Nations Environment Programme Ozone Secretariat 1997a, b).

Beijing Amendment and Adjustment: The 11th MOP held in Beijing in 1999a, b agreed that HCFCs and bromochloromethane (CH$_2$BrCl) would be included as controlled substances (United Nations Environment Programme Ozone Secretariat 1999a, b). It also banned the import and export of CH$_2$BrCl from and to non-Parties. Another Adjustment agreed in this MOP was the accelerated implementation of the CFCs, Halons, other fully halogenated CFCs and methyl bromide to the level of the basic domestic needs for developing countries (United Nations Environment Programme Ozone Secretariat 1999a, b).

Montreal Adjustment: No further changes to the obligations were decided for an 8-year period, but it was then decided at the 19th MOP in Montreal in 2007 to accelerate the control schedule of HCFCs for developing countries (United Nations Environment Programme Ozone Secretariat 2007).

Efforts for Further Amendments: Since MOP21 (2009), several countries have been proposing a freeze on HFC consumption and production. If adopted, such a freeze would be applicable to developed countries at first and then gradually to developing countries (United Nations Environment Programme 2009). This has been proposed because HFCs, originally developed as alternatives to ODS, have now been demonstrated to have a high global warming potential (United Nations Environment Programme 2015). Ongoing debate has discussed whether this falls under the mandate of the Montreal Protocol, as HFCs are not ODS. Either way, this fact only reinforces the importance of scientific findings and a holistic approach.

2.2.4   Satellite Instruments Used for Ozone Monitoring

2.2.4.1   Satellite Observations Before the Montreal Protocol

We seek to determine if satellite observations since the 1970s have had any influence on the choice of the substances specified in the Montreal Protocol and each amendment or adjustment thereof. The aim of this section is to study if there is a potential link to be found between satellite monitoring and the adoption and changing of international policy instruments.

Satellite measurements of total ozone started when the Infrared Interferometer Spectrometer (IRIS) and Backscatter Ultraviolet (BUV) instruments on Nimbus-4 were launched in 1970 (WMO 1981). Major instruments used in the 1970s included BUV, which continued to monitor for the next 7 years; the previously mentioned SBUV and TOMS, launched in 1978; TIROS Operational Vertical Sounder (TOVS) on TIROS-N; Limb Infrared Monitor of the Stratosphere (LIMS) on Nimbus-6; Stratospheric Aerosol and Gas Experiment (SAGE) on AEM-2; and the Upper Atmosphere Research Satellite (UARS). Most of the instruments/satellites were used to monitor the concentration of CO_2, O_3 and NO_2. SAGE (since 1979) and SAGE-II (since 1984) measured O_3, NO_2 and H_2O; and the Solar Mesospheric Explorer (SME) (launched in 1981) measured O_3, O_3 photodissociation rates, temperature, H_2O, thermal emissions, and NO_2 (WMO 1985a, b).

Until the late 1970s, atmospheric tides were just a theory to be verified by observation. The results of laboratory and ground-based experiments into the concentration of ozone in the middle and upper atmosphere also differed from one experimental technique to another, highlighting the need for direct, consistent, and comparable measurements. The scientific community anticipated the use of satellites for such measurements.[1] Stratospheric circulation and the exchange of constituents between the troposphere and the stratosphere started being measured in the late 1970s.

As early as 1981, the scientific community pointed out that newly introduced satellite observations considerably reduced the geographical limitations on measuring total ozone (WMO 1981). Remote sensing by satellite also contributed to recording large-scale, slowly varying disturbances, which provided a picture of the global mean distribution of ozone and its variations with a wide coverage (WMO 1981).

Climatology literature in the early 1980s stated that there was compelling evidence that the composition of the atmosphere was changing and some recent satellite monitoring, e.g., that of the vegetation index of plant activity in 1982, had been effective in precisely identifying the situation (WMO 1985a, b). It was emphasized that the continued development of baseline measurements for CO,

---

[1]It is stated in one WMO report that "[o]ne of the major difficulties encountered in the literature of estimating global total ozone variations is that there is no unique way to estimate such changes from the given ground-based data set" (WMO 1985a, b).

CH_4, N_2O, CO_2, and halocarbons were vital, and the already accomplished global coverage by Nimbus-7 TOMS ozone column measurements, Stratospheric and Mesospheric Sounder (SAMS) and LIMS data from Nimbus-7, and Stratospheric Sounding Unit (SSU) data from TIRON-N showed promise for the investigation of chemistry exchange processes (WMO 1985a, b). In addition to the SBUV and TOMS measurements, SAGE and SAGE-II measured O_3, NO_2, and H_2O; and the SME measured O_3, O_3 photodissociation rate, temperature, H_2O, thermal emission, and NO_2. For the future, it was proposed that "[a] satellite-borne CO sensor, operating for extended periods ($\sim$ years), could help enormously."

By the mid-1980s it was proven that the changing composition of the atmosphere influenced ozone depletion and that continued development of baseline measurements for CO, CH_4, N_2O, CO, and halocarbons (CH-11, CH-12, CCl_4, etc.) was vital to understand future ozone states.

While it may be said that satellite monitoring—in particular the global scale image of the ozone hole—played an important role in advancing ozone management policy to the extent that this helped the adoption of the Vienna Convention, it is fair to conclude that it was one of several techniques used to measure ozone concentration up until the mid-1980s. There is no evidence that satellite data by itself aided the decisions related to the concrete obligations of the Montreal Protocol (e.g., which specific substances to eliminate), though it can be said that scientific assessment by that time was a major source of information on which the drafting of the Montreal Protocol was based.

2.2.4.2 After the Montreal Protocol: Amendments and Adjustments

In 1998, just 1 year after the adoption of the Montreal Protocol, the WMO published another two-volume *Report of the International Ozone Trends Panel 1988* (WMO 1988a, b). The contents of this report differed greatly from the one published 3 years earlier. The most distinguished difference was the increased number of explanations of satellite-borne instruments and their use in measuring aerosol variations. It is even concluded in the report that: "either the ground-based or the satellite system could be assumed to be the standard, and the other system the one of uncertain quality to be tested against it" (WMO 1988a, b). Sensors such as SBUV are less susceptible to aerosol effects from ground-based natural phenomena, such as volcanic eruptions, and are regarded as highly efficient in detecting aerosol abundance (WMO 1988a, b). According to the report, there was extensive observation of trends in source gases including halocarbons such as CCl_3F, CCl_2F_2, CH_3CCl_3, CCl_4, other chlorocarbons, bromocarbon species, NO, CH_4 and trace gases influencing tropospheric ozone and hydroxyl radical concentrations. They were sometimes observed directly and other times indirectly via connected chemical processes. Some of the CFCs and HCFCs were added in the 1990 London Amendment and some of the CFCs and Halons were in the London Adjustment. Comparing the observed gases with those specified in the London Amendment, it can be said that satellite monitoring had an impact on ODS control decisions.

2.2.5   Circumstantial Evidence? Reports from SAP and ORM

The Scientific Assessment Panel (SAP), established in accordance with Art. 6 of the Montreal Protocol, assesses the status of the ozone layer and relevant atmospheric science issues. Technical and scientific assessments reported to the Parties at least every 4 years demonstrate:

i. the increased importance of scientific evidence for the amendment and adjustment of the Montreal Protocol; and,
ii. the increased importance of satellite monitoring among observations (WMO/UNEP 2016).

The relationships between scientific assessment and protocol amendment/ adjustment are seen in a chart from WMO/UNEP (2010), which is reproduced as Table 2.1. The contents of each Scientific Assessment of Stratospheric Ozone mostly correspond to the subsequent amendment/adjustment of the Montreal Protocol (WMO/UNEP 2014).

Table 2.1 Relationships between scientific assessment and protocol amendment/adjustment (WMO/UNEP 2010)

Year/policy process	Scientific assessment
1981	The stratosphere 1981: theory and measurements
1985 (Vienna convention)	Atmospheric ozone 1985
1987 (Montreal protocol)	
1988	International ozone trends panel report: 1988
1989	Scientific assessment of stratospheric ozone: 1989
1990	London amendment and adjustments
1991	Scientific assessment of ozone depletion: 1991
1992	Methyl bromide: its atmospheric science, technology, and economics (Assessment Supplement)
1992	Copenhagen amendment and adjustments
1994	Scientific assessment of ozone depletion: 1994
1995	Vienna adjustments
1997	Montreal amendment and adjustments
1998	Scientific assessment of ozone depletion: 1998
1999	Beijing amendment and adjustments
2002	Scientific assessment of ozone depletion: 2002
2006	Scientific assessment of ozone depletion: 2006
2007	Montreal adjustments
2010	Scientific assessment of ozone depletion: 2010
2014	Scientific assessment of ozone depletion: 2014

Likewise, as the chart in the executive summary of the *Scientific Assessment of Ozone Depletion: 2010* clearly shows, satellite observations are clearly of increased importance in the total assessment (WMO/UNEP 2010).

In addition to the SAP reports, Ozone Research Managers (ORM), established at the occasion of the first COP of the Vienna Convention in 1989, communicate scientific reports of ozone status every 3 years, 6 months before the respective COP (United Nations Environment Programme Ozone Secretariat 2016). Unlike the SAP reports, ORM reports include national reports. Examinations of each national report from 2002 (fifth meeting) to 2014 (ninth meeting) show that the States have increasingly relied on satellite Earth observation for monitoring the ozone situation.[2] The key sensors used by reporting states include TOMS on Nimbus-7, Meteor-3, ADEOS, and Earth Probe; SBUV on Nimbus-7, -9, -11, -14, -16, -17, -18, -19; SCIAMACHY on Envisat; OMI on EOS-Aura; TES on EOS-Aura; GOME-2 on Metop-A and Metop-B; SAM II on Nimbus-7; SAGE II on AEMB and ERBS; MLS on UARS and EOS-Aura; HALOE on UARS; POAM II on SPOT-3 and POAM III on SPOT-4; OSIRIS and SMR on Odin, SAGE-III on Meteor-3 M; MIPAS and GOMOS on Envisat; ACE on SCISAT; and HIRDLS on EOS-Aura (WMO 2016). Of these, the TOMS instrument is the most frequently cited. All of these sensors can measure O_3, NO_2, SO_2, HCHO, BrO and OCIO, and monitor various areas of the world. The missions are complementary and provide ozone measurement continuity from the initial TOMS/SBUV missions that had a significant impact on the early stages of decision-making in international ozone policy.

Drawing from all the WMO national reports from 2002–2014, the SAP reports, and other ozone monitoring techniques, it is clear that satellite monitoring has played an important role in the amendments and adjustments to the Montreal Protocol (Levelt et al. 2006; Gebhardt et al. 2014).

2.2.6 Tentative Conclusions and Way Forward

As a tentative conclusion, our comparison of the scientific and technical literature with the texts of the Montreal Protocol and subsequent amendments/adjustments suggests that the results of scientific assessments determine the ODS identified in the Montreal Protocol and its amendments/adjustments. Our research also suggests that there is a growing role for satellite observations. Therefore, it seems reasonable to conclude that satellite data, alongside other scientific evidence, have jointly determined the Montreal Protocol obligations.

---

[2]As archived satellite data has been used by reporting states, satellite instruments referred to by those states could be historical. Examples are the TOMS sensor on Nimbus-7 and Meteor-3 (1978–1994) as well as ADEOS and Earth Probe (1996–2006), which were mentioned even in the ninth ORM report in 2014 by Brazil, Chile, China, Indonesia, Mongolia, Norway, Turkey and USA (UNEP 2014).

Satellite Earth observations may have a long-term role to play in other policy studies. One such example could be the challenges posed by the changing Arctic environment.

Arctic sea ice needs to be monitored for safe ship navigation. Current global warming trends may be causing Arctic sea ice shrinkage, making arctic waterways more easily navigable without icebreaker ships. This could generate new issues related to the exploitation of natural resources in previously pristine areas. However, international and national laws in this area are inconsistent, and the Arctic Council has had to play a primary role in governance in order to establish international policies in areas where natural conditions are rapidly changing. The eight Member States of the Arctic Council are the coastal states of the Arctic. The Member States decide most of the rules. Other states have interests in this area (such as Japan, China, and Korea), but they are only allowed to become Observer States and cannot participate directly in the rule-making process. In this situation, a state with highly advanced technology can contribute to policymaking by providing the scientific data necessary for Arctic research.

Japan carries out many research projects, such as the Green Network of Excellence (GRENE) Arctic Climate Change Research Project and the Arctic Challenge for Sustainability (ArCS) project led by the Ministry of Education, Culture, Sports, Science and Technology (MEXT).

Japan operates two satellites capable of supporting this issue:

i. Ibuki, the Greenhouse gases Observing SATellite (GOSAT) led by JAXA, the Ministry of the Environment (MOE), and the National Institute for Environmental Studies (NIES), is the first satellite designed specifically to observe CO_2, CH_4, and other greenhouse gases to monitor their status/inventory; and,

ii. GCOM-W, led by JAXA, provides unique information for monitoring the size of arctic sea ice.

Another powerful tool would be precise weather forecasting in the Arctic, whereby multiple satellites observe water vapor, clouds, snow, and raindrops with a temporal–spatial resolution of about 1 h and in the order of 1–2 km (not only horizontal but also vertical) using Terahertz (THz) microwave passive sensors. THz technology would enable the use of mini-satellites for these multi-satellite observations. The data resulting from Japanese research would be recognized as important scientific knowledge and might contribute to the protection and effective use of the Arctic.

2.3 A Quantitative Approach for Linking Policy and Satellite Earth Observation Using Text Mining Techniques

2.3.1 Introduction

At first glance, policy or decision-making and satellite Earth observation belong to entirely different fields. However, there is a commonality if we focus on the reports and publications that contain information from both of these areas. This section describes a text mining technique to quantify possible linkages between decision-making and satellite Earth observation.

Text mining retrieves high quality information from text data using statistical and mathematical techniques. In the mid-1980s, labor-intensive manual work was the primary approach to text mining. Advances in computing technology over the past few decades have allowed the automation of text mining methods and the development of many useful computer-coded tools. Text mining is now widely applied in many fields, not only in science but also in marketing, for example for the design of new products using information from customer surveys. Natural language processing (NLP) techniques are essential. NLP is the analysis of human language so that computers can understand natural languages as human do. With the assistance of human experts, NLP techniques are capable of providing a way to verify the detected relationship across different text collections.

Below we describe an investigation into the role of satellite Earth observation in the decision-making process for the Montreal Protocol. A quantitative linkage between the decision-making process and satellite Earth observation is then described.

2.3.2   Role of Satellite Earth Observation in Policy Decisions for the Montreal Protocol

The reports of the MOP to the Montreal Protocol were used as the documents that represent the policy decision-making process. As described above, the Montreal Protocol has been updated through several amendments and adjustments. These amendments and adjustments are discussed and determined by the MOP, whose reports have been published every year from 1990.

One basic and commonly used approach in text mining involves counting the number of times a word occurs. We performed a morphological analysis that assigns a part of speech for all words in 25 MOP reports between 1990 and 2014. Only nouns, verbs, and adjectives were extracted from the MOP reports because other less meaningful words, such as articles "the" or "a", function as noise in this analysis. We used the Stanford Log-Linear Part-Of-Speech Tagger (Toutanova and Manning 2000; Toutanova et al. 2003) for the morphological analysis. Figure 2.2 shows an example result of the morphological analysis in this study. To unify words with different notations, such as "satellite" and "satellites," plural nouns were converted to the singular. Verbs were converted to the base form. Linking verbs, such as "am", "are", and "be", were omitted from the documents. Abbreviations and fully spelled words were also unified into abbreviations. For example, "hy-drofluorocarbon (HFC)" was converted to "hfc". All capital letters in the documents were converted to lower case.

Recognizing that the Ozone Research Managers at their seventh
meeting emphasized the discussion of satellite research and
monitoring owing to concerns among scientists that there could be
serious gaps in satellite monitoring and associated data when current
generation satellites and associated instruments reach the end of their
useful lives in the next few years,

 lemmatization (+stop word removal)

recognize orm seventh meeting emphasize discussion
satellite research monitoring owing concern scientist serious
gap satellite monitoring associate data current generation
satellite associate instrument reach end useful life next few
year

Fig. 2.2 An example of the morphological analysis in this study. The original sentence is quoted
from decision VCVIII/2: recommendations Adopted by the Ozone Research Managers at their
seventh meeting (United Nations Environment Programme Ozone Secretariat 2008)

We focused on the word "satellite" to investigate how satellite Earth observation
is dealt with in the MOP. The word "satellite" appeared 19 times in all of the 25
MOP reports. All of the words "satellite" were used in relation to satellite Earth
observation in the MOP reports, so we may take the word "satellite" as being
representative of the topic of satellite Earth observation. Figure 2.3 shows the
number of occurrences of the word "satellite" in the MOP reports over time. The
results showed that the word "satellite" appeared 13 times in 2008, so we focused
on this year.

To investigate the topic in which the word "satellite" is used, we selected the
sentences that include the word "satellite" and the sentences immediately before
and after. We then extracted the words that appeared more than five times in the
selected sentences. To extract significant words from those in the selected sentences, we introduced a term frequency-inverse document frequency (*tf-idf*) method
to quantify the significance of a word. The basic goal of *tf-idf* is to quantify the
significance of each word in a document. In this study, we employed a *tf-idf* defined
by Gamon et al. (2005) as follows. Here w and d represent the indexes of the word
and document.

$$tf\text{-}idf_{w,d} = \begin{cases} \{1 + \log(tf_{w,d})\} \times log(N_{doc}/df_w), & tf_{w,d} \geq 1 \\ 0, & tf_{w,d} = 0 \end{cases} \quad (2.1)$$

A detailed explanation of *tf-idf* is described below. The simplest approach when
assigning the significance of the word w in document d is to count the number of
occurrences, which is called term frequency ($tf_{w,d}$). In this equation, $tf_{w,d}$ is modified to be $1 + \log(tf_{w,d})$. However, the raw *tf* value is not enough to quantify the
significance of the words because it is impossible to distinguish those words that

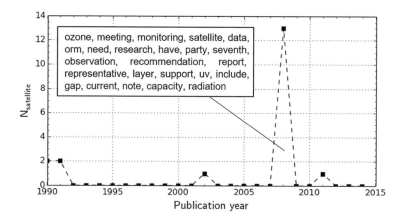

Fig. 2.3 Number of occurrences of the word "satellite" ($N_{satellite}$) in the MOP reports from 1990 to 2014. The words in the *square box* are those that appear more than five times in the selected sentences around the word "satellite"

appear in all documents from those that appear only in a few. If the *tf* values of these words are equal, the significance should be different. For example, the word "meeting" appears many times in the MOP reports but the significance of the word "meeting" should be low. To solve this problem, the idea of document frequency (*df*) is introduced. The document frequency is defined as the number of documents that contain the word *w*. The significance of a word should be high when the *df* value is low, thus the inverse document frequency (*idf*) of a word scaled with the total number of documents N_{doc} is introduced. In this study, the value of *idf* for the word *w* is calculated by $\log(N_{doc}/df_w)$. The value of *tf-idf*$_{w,d}$ is calculated by $tf_{w,d} \times idf_w$.

Here each MOP report is treated as one document. The five most significant words determined by *tf-idf* were "satellite", "orm", "gap", "seventh" and "observation". The word "orm" means the Ozone Research Managers (ORM), which include government atmospheric research managers and government managers of research related to health and environmental effects of ozone modifications. The ORM meeting reviews ongoing national and international research and makes recommendations for future research and expanded co-operation between researchers in developed and developing countries for consideration by the Conference of the Parties to the Vienna Convention (Table 2.2).

In the seventh ORM meeting held in 2008, a problem with the NASA Aura satellite mission was discussed. The Aura satellite is equipped with four instruments to observe ozone and ozone-related species in the stratosphere and troposphere. The Aura satellite and its predecessor NASA satellites, such as the Upper Atmosphere Research Satellite (UARS) and Nimbus series, have been continuously observing the Earth's atmosphere. The problem was that some of the instruments might have stopped operating anywhere between 2008 and 2010, resulting in a serious gap in the long-term monitoring of ozone since the 1970s. Fortunately, the Aura satellite is

Table 2.2 Ranking of words
with high *tf-idf* values

Word	*tf–idf*	N_{word}
"Satellite"	5.738	13
"Orm"	5.515	11
"Gap"	2.424	8
"Seventh"	1.846	8
"Observation"	1.783	8

orm ozone research managers, N_{word} number of word occurrences

still operational as of July 2016 and other agencies' satellites, such as Odin and SCISAT, also continue to monitor atmospheric ozone. As a consequence, there has been no serious gap in the long-term satellite monitoring of ozone. We emphasize that the MOP in 2008 also raised this problem and mentioned the importance of long-term monitoring using satellites. It is clear that satellite Earth observation plays a role in policy decisions by providing data on long-term trends of atmospheric substances.

2.3.3 Quantifying the Correlation Between Policy and Satellite Earth Observation

The Montreal Protocol has been updated through amendments and adjustments. Any discussion of amendments and adjustments should be based on scientific evidence, such as long-term trends of ozone and species related to ozone. We therefore have focused on the discussion phase for amendments and adjustments to quantify the correlation between policy, or decision-making, and satellite Earth observation.

The MOP has recently focused on hydrofluorocarbons (HFCs), a significant non-ODS alternative to CFCs, which have, however, now been shown to have significant global warming potential. Since the latter half of the 1990s, global warming and climate change have joined ozone depletion as major issues on the global environmental agenda. In 1997, the Kyoto Protocol to the UN Framework Convention on Climate Change (UNFCCC) was adopted and entered into force in 2005. The Protocol aims to provide specific goals for international efforts to reduce the emission of greenhouse gases. The phasing out of HFC production and consumption was first proposed by several countries at MOP21 (2009).

Figure 2.4a shows the number of occurrences of the word "hfc" (N_{HFC}) in the MOP reports over time. The adoption of the Kyoto Protocol in 1997 triggered a rapid increase in N_{HFC} in 1998. After the phasing out of HFCs was proposed in 2009, N_{HFC} rose to over 150 in 2014. The increasing ratio (*IR*) of the word "hfc" was the largest among the substances appearing in the MOP reports.

Table 2.3 shows the five words with the highest *IR* values in the MOP reports. In this study, the value of *IR* for the word w (IR_w) was defined as follows. N_{doc} is the

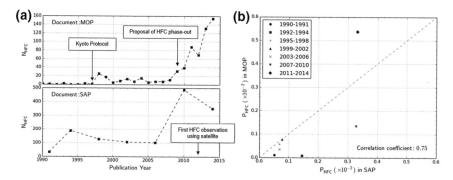

Fig. 2.4 **a** Time trends for the number of occurrences of the word "hfc" (N_{HFC}) in the MOP (*upper panel*) and SAP reports (*lower panel*) from 1990 to 2014, **b** Scatter plot of the probabilities of occurrences (P_{HFC}) in the MOP and SAP reports. The MOP probabilities are averaged for a period corresponding to the time interval of the SAP report publications. The correlation coefficient is estimated to be 0.75

Table 2.3 Ranking of words with high *IR* values among the words of substances

Word	Full Word	IR
"hfc"	Hydrofluorocarbon	0.409
"ghg"	Greenhouse gas	0.258
"hfo"	Hydrofluoroolefin	0.239
"h2o"	Water vapor	0.231
"co2"	Carbon dioxide	0.175

total number of documents. N_m is the number of documents to calculate the mean standard of the number of occurrences and was set to four in this analysis, corresponding to the time interval of the publication of the Scientific Assessment of Ozone Depletion prepared by the SAP.

$$IR_w = \text{Mean}\left(ir_{w,N_m}, \ldots, ir_{w,N_{doc}}\right), ir_{w,d} = \frac{N_{w,d} - \text{Mean}\left(N_{w,d-N_m}, \ldots, N_{w,d}\right)}{\text{Mean}\left(N_{w,d-N_m}, \ldots, N_{w,d}\right)} \quad (2.2)$$

The SAP report is used to represent scientific research publications in this study. The SAP reports are prepared every 3–4 years based on a number of scientific papers, pursuant to Article 6 of the Montreal Protocol. The Panel consists of hundreds of top scientists and investigates the status of the depletion of the ozone layer and relevant atmospheric scientific issues. The lower panel of Fig. 2.4a shows the time trend of N_{HFC} in the SAP reports published from 1991 to 2014. In 2010, N_{HFC} increased to about 500, which is consistent in terms of time with the proposal of HFC phase out at MOP21 in 2009. Figure 2.4b shows the correlation between the probabilities of occurrences of the word "hfc" (P_{HFC}) in the MOP and SAP reports. The value of P_{HFC} is calculated by N_{HFC} divided by the total number of words (N_{All}). The value of P_{HFC} was averaged during the time interval of the SAP report publications. This figure shows a positive correlation of P_{HFC} between the MOP and SAP reports, with a correlation coefficient of 0.75. Although we could not identify from this analysis whether science drives policy or vice versa, this does demonstrate a certain connection between policy and science.

As described in Sect. 2.3.2, the role of satellite Earth observation is to provide decision-makers information on the long-term trends of atmospheric substances. Satellite observations of HFCs were first reported in 2012 using the Atmospheric Chemistry Experiment Fourier Transform Spectrometer (ACE-FTS) onboard the SCISAT-1 satellite (Harrison et al. 2012). The long-term trend of HFCs derived from satellite Earth observations might accelerate further amendments covering HFC production and consumption.

2.3.4 Conclusion

The correlation between policy and satellite Earth observation was investigated for the Montreal Protocol using basic text mining techniques. We utilized several methods and found that correlations by trend analysis with temporal sequences were the best approach to assessing satellite observations and the Montreal Protocol. Tracing the word "satellite" in the MOP reports, our analysis suggested that satellite Earth observation plays a role in policy through the provision of data on long-term trends in atmospheric substances. A certain correlation (correlation coefficient of 0.75) was shown using the word "hfc" for temporal trends in the text from MOP and SAP reports.

To our knowledge, this is the first investigation of whether satellite Earth observations have an impact on policy using text mining techniques. More in-depth investigations are needed to provide a more detailed understanding.

2.4 Conclusions from the Study of the Policy and Earth Observation Innovation Cycle

The research project entitled "Policy and Earth Observation Innovation Cycle" began in November 2013. The objective was to identify why satellite Earth observations are needed in the context of global environmental policymaking.

Satellite Earth observations provide useful information for agriculture, resource exploration, industry, and numerous other applications. Meteorological satellites are perhaps the most prominent, facilitating weather forecasting.

Questions remain over the use of satellites for environmental monitoring—who are the public users other than research scientists and are the data obtained providing societal benefits consistent with the taxpayers' investment? It appears that a quantitative answer has yet to be developed as a rationale to support the use of satellites for environmental observations.

The functions of satellite observations can be categorized according to their spatiotemporal resolutions. Figure 2.5 presents a summary of satellite usage by spatiotemporal scale. It is clear that environmental satellite observations are currently used for long-term trends and global problems. We therefore analyzed global policy in our study.

The purpose of the PEOIC project was to find a method for the quantitative evaluation of the impact of satellite Earth observations on policy. Planning for satellite observations to address environmental issues—such as ozone depletion, global warming, and air quality—has always been part of policy, such as the Montreal Protocol, the UNFCCC and the Climate and Clean Air Coalition. We have attempted to provide quantitative evidence of the effect of satellite observations on policy by using a mathematical method.

The process of the project was:

1. Define the general system of the PEOIC as shown in Fig. 2.1 in Sect. 2.1 (led by Yasuko Kasai with the assistance of Prof. Yoshifumi Yasuoka).
2. Perform an intensive survey of past evaluation methods for several satellite observations (led by Masami Onoda). The survey identified the limitations of assessing the impact of satellite data on policy using document analysis. Only a limited number of sample documents can be used for such analysis. It was also found that even if it is possible to analyze the relationship between satellite data and science, the relationship between science and policy is not straightforward and is difficult to quantify due to the many complicated factors involved.
3. Select the target policy (all team members). We selected an example of a previously successful policy, namely the Montreal Protocol. Satellite observation is

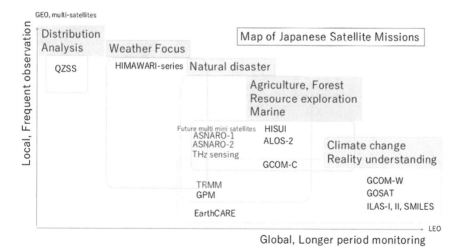

Fig. 2.5 Overview of current Japanese satellite missions according to the spatiotemporal scale of their observations

the only feasible way to monitor the overall size of the ozone hole in the polar regions, and provides scientific evidence of ozone layer recovery.

4. Define the flow between the Montreal Protocol and the scientific data, including satellite Earth observations (all team members). We confirmed that scientific observations, including satellite measurements, have different roles in the policymaking process. There was an increase in the number of scientific papers and satellite operations (as shown in Fig. 2.6) after the Montreal Protocol was agreed in 1987 in a bid to understand the reality and mechanisms of ozone depletion. Our study demonstrated that the agreement of the Montreal Protocol in 1987 resulted in increased satellite launches.

5. Determination of the documentation used in policy and scientific research (all members). This task was very important for our approach and it took more than one year to define the documentation. Figure 2.7 demonstrates the scope of this task with an organizational map of the Vienna Convention and Montreal Protocol (prepared by Setsuko Aoki). WMO/UNEP Scientific Assessments of Ozone Depletion were also used.

6. In-depth evaluation of the policy and scientific documentation, as summarized in Sect. 2.2 (Setsuko Aoki). We conclude that the results from scientific assessments determine the ODS to be eliminated in the Montreal Protocol and its amendments/adjustments. It seems that satellite data alongside other scientific evidence jointly determined the obligations of the Montreal Protocol system.

7. Perform data mining to determine the correlation between scientific input and MOP decisions, as shown in Sect. 2.3 (Tomohiro Sato with Akiko Aizawa, Masami Onoda, Setsuko Aoki, Yasuko Kasai). We tried several methods and found the correlations from trend analysis with temporal sequences to be the

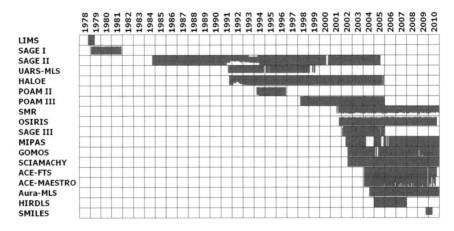

Fig. 2.6 Figure 2.1 from Tegtmeier et al. (2013). The number of limb-sounding satellite instruments used to observe ozone and related species increased after the agreement of the Montreal Protocol in 1987. Note that at that time it took about 10–15 years to develop a new satellite

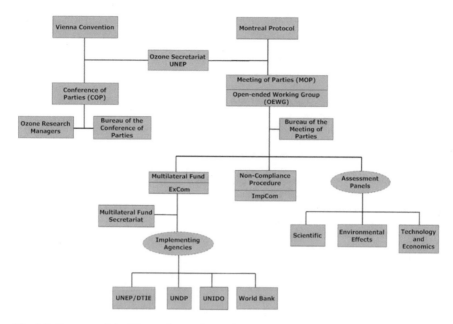

Fig. 2.7 Structure of the Vienna Convention and Montreal Protocol, produced by Setsuko Aoki

best. By tracing the word "satellite" in the MOP reports, the results suggested that satellite Earth observation played a role in policy by providing data on long-term trends in atmospheric substances. A certain correlation (coefficient of 0.75) was shown using the word "hfc" for temporal trends in the text from the MOP and SAP reports.

In conclusion, we have found that satellite monitoring has certainly played an important role in amending/adjusting the Montreal Protocol.

Acknowledgements We are grateful to Prof. Kazuo Matsushita of IGES for guiding the team in initiating and carrying out the project.

References

Booz & Company (2011) Cost-Benefit Analysis for GMES

Future Earth (2016) Home | Future Earth. http://www.futureearth.org/. Accessed 24 Nov 2016

Gamon M et al (2005) Pulse: Mining customer opinions from free text. In: Proceedings of the international symposium on intelligent data analysis, pp 121–132

Gebhardt C et al (2014) Stratospheric ozone trends and variability as seen by SCIAMACHY from 2002 to 2012. Atmos Chem Phys 14(2):831–846

Harrison JJ et al (2012) First remote sensing observations of trifluoromethane (HFC-23) in the upper troposphere and lower stratosphere. J Geophys Res Atmos p 117(D5)

IPCC (2001) Working Report, I. Scientific Basis

Levelt PF et al (2006) Science objectives of the ozone monitoring instrument. IEEE Trans Geosci Remote Sens 44:1199–1206

Macauley MK (2009) Earth observations in social science research for management of natural resources and the environment. Resources for the Future, Washington, DC

Manning CD et al (2008) Introduction to information retrieval. Cambridge University Press, New York

Molina MJ, Rowland FS (1974) Stratospheric sink for chlorofluoromethanes: chlorine atom-catalysed destruction of ozone. Nature 249:810–812

NASA Ozone Watch (2013) Ozone hole watch: facts about ozone hole history. http://ozonewatch. gsfc.nasa.gov/facts/history.html. Accessed 24 Apr 2016

NRC (2007) Earth science and applications from space: national imperatives for the next decade and beyond. National Academies Press, Washington, D.C., USA

PricewaterhouseCoopers LLP (2006) Main report socio-economic benefits analysis of GMES

Tegtmeier S et al (2013) SPARC Data Initiative: a comparison of ozone climatologies from international satellite limb sounders. J Geophys Res Atmos 118(21):12,229–12,247

Toutanova K, Manning CD (2000) Enriching the knowledge sources used in a maximum entropy part-of-speech tagger. http://nlp.stanford.edu/manning/papers/emnlp2000.pdf. Accessed 7 Dec 2016

Toutanova K et al (2003) Feature-rich part-of-speech tagging with a cyclic dependency network. In: Proceedings of HLT-NAACL, Edmonton, pp 252–259

United Nations (1985) Vienna convention for the protection of the ozone layer. United Nations Treaty Series, Vienna

United Nations (1987) Montreal protocol on substances that deplete the ozone layer. United Nations Treaty Series, Montreal

United Nations Environment Programme (2009) UNEP/OzL. Pro. 21/8

United Nations Environment Programme (2010) Key achievements of the Montreal protocol to date. http://ozone.unep.org/Publications/MP_Key_Achievements-E.pdf. Accessed 24 Apr 2016

United Nations Environment Programme (2014) Report of the ninth meeting of the ozone research managers of the parties to the Vienna convention for the protection of the ozone layer

United Nations Environment Programme (2015) UNEP/OzL. Pro. 27/13

United Nations Environment Programme Ozone Secretariat (1990) The London Amendment (1990): the amendment to the Montreal protocol agreed by the second meeting of the parties

United Nations Environment Programme Ozone Secretariat (1992) The Copenhagen Amendment (1992): the amendment to the Montreal protocol agreed by the fourth meeting of the parties

United Nations Environment Programme Ozone Secretariat (1996a) Adjustments agreed at the seventh meeting of the parties relating to controlled substances in annex B. http://ozone.unep.org/ en/handbook-montreal-protocol-substances-deplete-ozone-layer/27599. Accessed 14 Apr 2016

United Nations Environment Programme Ozone Secretariat (1996b) Adjustments agreed at the seventh meeting of the parties relating to controlled substances in annexes C and E. http://ozone.unep.org/en/handbook-montreal-protocol-substances-deplete-ozone-layer/27600. Accessed 14 Apr 2016

United Nations Environment Programme Ozone Secretariat (1997a) Adjustments agreed in 1997 in Montreal at the ninth meeting of the parties. http://ozone.unep.org/en/handbook-montreal-protocol-substances-deplete-ozone-layer/27601. Accessed 14 Apr 2016

United Nations Environment Programme Ozone Secretariat (1997b) The Montreal Amendment (1997): The amendment to the Montreal protocol agreed by the ninth meeting of the parties. http://ozone.unep.org/en/handbook-montreal-protocol-substances-deplete-ozone-layer/27611. Accessed 4 Apr 2016

United Nations Environment Programme Ozone Secretariat (1999a) Adjustments agreed in 1999 in Beijing at the eleventh meeting of the parties. http://ozone.unep.org/en/handbook-montreal-protocol-substances-deplete-ozone-layer/27604. Accessed 14 Apr 2016

United Nations Environment Programme Ozone Secretariat (1999b) The Beijing Amendment (1999): the amendment to the Montreal protocol agreed by the eleventh meeting of the parties.

http://ozone.unep.org/en/handbook-montreal-protocol-substances-deplete-ozone-layer/27612. Accessed 14 Apr 2016

United Nations Environment Programme Ozone Secretariat (2007) Adjustments agreed in 2007 in Montreal at the nineteenth meeting of the parties. http://ozone.unep.org/en/handbook-montreal-protocol-substances-deplete-ozone-layer/27607. Accessed 14 Apr 2016

United Nations Environment Programme Ozone Secretariat (2008) Decision VCVIII/2: recommendations adopted by the ozone research managers at their seventh meeting. http://ozone.unep.org/en/handbook-vienna-convention-protection-ozone-layer/2344. Accessed 4 Apr 2016

United Nations Environment Programme Ozone Secretariat (2016) Institutions | Ozone Secretariat. http://ozone.unep.org/en/institutions. Accessed 25 Apr 2016

Williamson RA et al (2002) The socio-economic value of improved weather and climate information. Space Policy Institute, The George Washington University, Washington, D.C

WMO (1981) The stratosphere 1981: theory and measurements. Geneva, Switzerland

WMO (1985a) Atmospheric ozone 1985: assessment of our understanding of the processes controlling its present distribution and change, vol. 1. Geneva, Switzerland

WMO (1985b) Atmospheric ozone 1985: assessment of our understanding of the processes controlling its present distribution and change, vol. 2. Geneva, Switzerland

WMO (1988a) Report of the international ozone trends Panel-1988, vol.1. Geneva, Switzerland

WMO (1988b) Report of the international ozone trends Panel-1988, vol. 2. Geneva, Switzerland

WMO (2016) WMO global ozone research and monitoring project reports. https://www.wmo.int/pages/prog/arep/gaw/ozone_reports.html. Accessed 25 Apr 2016

WMO/UNEP (2010) Scientific assessment of ozone depletion: 2010. https://www.wmo.int/pages/prog/arep/gaw/ozone_2010/documents/Ozone-Assessment-2010-complete.pdf. Accessed 7 Dec 2016

WMO/UNEP (2014) Assessment for decision-makers: scientific assessment of ozone depletion: 2014. http://www.wmo.int/pages/prog/arep/gaw/ozone_2014/documents/ADM_2014Ozone Assessment_Final.pdf. Accessed 7 Dec 2016

WMO/UNEP (2016) Quadrennial assessment and progress reports. http://ozone.unep.org/en/assessment-panels/scientific-assessment-panel. Accessed 5 May 2016

Author Biographies

Yasuko Kasai is an executive researcher at the Terahertz Technology Research Center and Big Data Integration Research Center within the National Institute of Information and Communications Technology (NICT), professor at the Tokyo Institute of Technology, and an officer of the Ministry of Internal Affairs and Communications. She previously served as special postdoctoral researcher (SPDR) at RIKEN. She received her Ph. D. from Tokyo Institute of Technology in 1995. Prof. Kasai is a member of the COSPAR Scientific Commission and a commissioner of the International Radiation Commission (IRC). Her research area is terahertz satellite remote sensing of the Earth and other planets.

Setsuko Aoki is professor of international law in the Faculty of Policy Management at Keio University, Japan. Her previous positions include being an associate professor at Keio and associate professor of the School of Social Science, National Defense Academy. She received her Doctor of Civil Law (D.C. L.) from the Institute of Air and Space Law, Faculty of Law, McGill University, Canada, in June 1993; her Master of Laws (LL.M.) from the Graduate School of Law, Keio University, Japan, in March 1985; and a Bachelor of Civil Law (B.C.L.) from the Faculty of Law, Keio University, Japan, in March 1983.

Professor Aoki has served as a member of the Committee on National Space Policy (Cabinet Office) since July 2012, as a legal advisor to the Ministry of Foreign Affairs, and on the Legal Subcommittee of the Committee on the Peaceful Uses of Outer Space (COPUOS) since March 2002. Her area of research is international law and space law.

Akiko Aizawa graduated from the Department of Electronics at The University of Tokyo in 1985 and completed her doctoral studies in electrical engineering in 1990. She was a visiting researcher at the University of Illinois at Urbana-Champaign from 1990 to 1992. At present, she is a professor at the National Institute of Informatics (NII) and also an adjunct professor at the Graduate School of Information Science and Technology (IST) at The University of Tokyo. Her research interests include text-based content and media processing, statistical text analysis, linguistic resource construction, and corpus-based knowledge acquisition.

Akiko Okamatsu is a professor of international law in the Faculty of Sustainability Studies at Hosei University, Japan. She graduated from Jouchi (Sophia) University in Tokyo. She is a member of the Compliance Group of the 1996 Protocol to the Convention on the Prevention of Marine Pollution by Dumping of Wastes and Other Matter 1972; special advisor to the Ministry of Environment, Japan; and guest researcher at the Research Office of the Standing Committee on Foreign Affairs and Defense, House of Councillors, Japan. She began her career as a research associate at Jouchi University, Faculty of Law, before becoming a postdoctoral fellow at the National Institute for Environmental Studies (NIES) and later assistant professor at Shobi University. She also served as a visiting scholar at Harvard Law School from 2013 to 2014.

Tomohiro Sato is a researcher at the National Institute of Information and Communications Technology (NICT), Japan. His previous positions include teacher of science at Ichikawa Junior High School and High School and research fellow at the Japan Society for the Promotion of Science. His Doctorate of Science, M.Sc., and B.Sc. degrees are from the Tokyo Institute of Technology, Japan. His research interests are Earth observation, atmospheric science, remote sensing, and text mining.

Masami Onoda is currently the U.S. and multilateral relations interface at the International Relations and Research Department of the Japan Aerospace Exploration Agency (JAXA). As an academic, she is fellow of the Institute of Global Environmental Strategies (IGES) and she is also engaged in the private sector as an advisor to the Singapore-based space debris start-up Astroscale Pte. Ltd. since its foundation in 2013. From 2009 to 2012, Dr. Onoda was a scientific and technical officer at the intergovernmental Group on Earth Observations (GEO) Secretariat in Geneva, Switzerland. From 2003 to 2008, while pursuing her graduate studies, she was invited to the JAXA Kansai Satellite Office in Higashiosaka as a space technology coordinator to support technology transfer to SMEs for the small satellite project SOHLA-1. From 1999 to 2003, she worked in the field of Earth observations at JAXA (then NASDA), serving on the Secretariat of the Committee on Earth Observation Satellites (CEOS). In 1999, she was seconded to the UN Office for Outer Space Affairs (UNOOSA) for the organization of the UNISPACE III conference. She holds a Ph.D. in global environmental studies (2009) and a master's degree in environmental management (2005), both from the Kyoto University Graduate School of Global Environmental Studies. Her undergraduate degree is in international relations from The University of Tokyo.

Brian Alan Johnson received his Ph.D. in geosciences from Florida Atlantic University in 2012 and his M.A. in geography from the same university in 2007. He is now a researcher at the Institute for Global Environmental Strategies (IGES) in Hayama, Japan. His research interests in remote sensing are related to land use/land cover mapping, change detection, and data fusion.

Part III
National and Regional Experiences
(U.S., Europe, and Asia)

Chapter 3
Innovation in Earth Observations as a National Strategic Investment: The Experience of the U.S.

Masami Onoda and Molly Macauley

3.1 U.S. Earth Observation Strategy

Earth observations can provide actionable science and information for policy through projections and forecasts; near real-time management or identification of deviations in long-term trends; and improved spatial, spectral, or temporal resolution (and better characterization of uncertainty) in data, observations, and models. Earth observations are regularly used by the media and interested members of the public, with the capacity to visualize large amounts of quantitative data using today's technology infrastructure.

Molly Macauley—Deceased

Dr. Molly Macauley was invited as a member of the PEOIC International Advisory Board and joined the workshop in Tokyo in November 2015. In a tragic incident Dr. Macauley passed away on July 8, 2016, her contribution for this book yet to be completed. The following article was written by Dr. Masami Onoda on Dr. Macauley's behalf, with permission from her institute, based on Dr. Macauley's presentation at the Tokyo workshop. Molly's dedication and expertise with which she made such an enormous contribution to the study of the benefits of space will never be forgotten, including this valuable piece of work with our group. It has truly been an honor and privilege to work with her.

M. Onoda (✉)
International Relations and Research Department, Japan Aerospace Exploration Agency,
Ochanomizu Sola City, 4-6, Kanda Surugadai, Chiyoda-Ku, Tokyo 101-8008, Japan
e-mail: onoda.masami@jaxa.jp

M. Macauley
Resources for the Future (RFF), 1616 P St. NW, Suite 600, Washington, DC 20036, USA

© The Author(s) 2017
M. Onoda and O.R. Young (eds.), *Satellite Earth Observations and Their Impact on Society and Policy*, DOI 10.1007/978-981-10-3713-9_3

3.1.1 National Space Policy of the U.S.

The U.S. policy for Earth observation is set out in the National Space Policy of the United States of America (Executive Office of the President 2010), and includes an emphasis on expanding and identifying international cooperation opportunities. The Policy states in its introduction that: "the United States hereby renews its pledge of cooperation in the belief that with strengthened international collaboration and reinvigorated U.S. leadership, all nations and peoples—space-faring and space-benefiting—will find their horizons broadened, their knowledge enhanced, and their lives greatly improved." The goals of the Policy include: "expand international cooperation on mutually beneficial space activities to: broaden and extend the benefits of space; further the peaceful use of space; and enhance collection and partnership in sharing of space-derived information." In the section on international cooperation, the Policy advocates open access to Earth observation data by promoting "the adoption of policies internationally that facilitate full, open, and timely access to government environmental data." The Policy further states that "departments and agencies shall identify potential areas for international cooperation that may include, but are not limited to: space science; space exploration, including human space flight activities; space nuclear power to support space science and exploration; space transportation; space surveillance for debris monitoring and awareness; missile warning; Earth science and observation; environmental monitoring; satellite communications; global navigation satellite systems (GNSS); geospatial information products and services; disaster mitigation and relief; search and rescue; use of space for maritime domain awareness; and long-term preservation of the space environment for human activity and use." In the section *Environmental Earth Observation and Weather*, the Policy stresses the use of international partnerships to help sustain and enhance weather, climate, ocean, and coastal observation from space.

3.1.2 National Plan for Civil Earth Observations

The U.S. National Plan for Civil Earth Observations (National Science and Technology Council, Executive Office of the President 2014) defines a new framework for constructing a balanced portfolio of Earth observations and observing systems. This framework classifies Earth observation activities according to two broad categories—*sustained* and *experimental*—based on the duration of the anticipated Federal commitment:

- Sustained observations are defined as measurements taken routinely that Federal agencies are committed to monitoring on an ongoing basis, generally for 7 years or more. These measurements can be for public services or for Earth system research in the public interest.

- Experimental observations are defined as measurements taken for a limited observing period, generally 7 years or less, that Federal agencies are committed to monitoring for research and development purposes. These measurements serve to advance human knowledge, explore technical innovation, and improve services, and in many cases may be first-of-their-kind Earth observations.

Based on this framework and the results of the Earth Observation Assessment (EOA),[1] the National Plan establishes the following rank-ordered priorities:

1. Continuity of sustained observations for public services
2. Continuity of sustained observations for Earth system research
3. Continued investment in experimental observations
4. Planned improvements to sustained observation networks and surveys for all observation categories
5. Continuity of, and improvements to, a rigorous assessment and prioritization process.

Action 7[2] of the Plan states: "Maintain and Strengthen International Collaboration. The global nature of many Earth observations and the value of these observations to U.S. Government decision-makers require U.S. agencies to carry out their missions through collaboration with foreign agencies, international organizations, and standards/coordination groups. Through international collaboration, U.S. agencies leverage foreign data and scientific expertise to improve their understanding of remote areas, such as the open ocean and polar regions, and to characterize global atmospheric, oceanic, and terrestrial phenomena. In addition, collaboration with international partners helps to minimize unnecessary redundancy in the collection of Earth observations, and ensures the effective use of limited resources. U.S. agencies also work closely with the Department of State and other agencies to provide associated scientific and technical support for U.S. foreign policy, security, economic, and environmental interests."

3.1.3   A Plan for a U.S. National Land Imaging Program

For land observations, a plan for a U.S. national land imaging program (Future of Land Imaging Interagency Working Group, Executive Office of the President 2007) was initiated as an "effort to develop a long-term plan to achieve technical, financial, and managerial capability for operational land imaging in accord with the goals and objectives of the U.S. Integrated Earth Observation System." Noting that the "importance of this imagery to the Nation requires a more sustainable effort to ensure that land imaging data are available far into the future," it was developed as

---

[1]See Chap. 2 of the U.S. National Strategy for Civil Earth Observations (National Science and Technology Council, Executive Office of the President 2013).

[2]See Sect 4.2.7 of the National Plan for Civil Earth Observations (National Science and Technology Council, Executive Office of the President 2014)

"a plan that identifies options for future civil land imaging data acquisition, establishes an implementation strategy, and recommends a governance and management structure to ensure that future U.S. land imaging needs will be met." "It presents a set of policy recommendations to achieve a stable and sustainable U.S. operational space-based land imaging capability and to ensure continued U.S. scientific, technological, and policy leadership in civil land imaging and the scientific disciplines it supports."

3.1.4 Decadal Survey for Earth Science and Applications from Space

The so-called "Decadal Survey" (National Academy of Sciences 2007) is based on the request of NASA, the National Oceanic and Atmospheric Administration (NOAA)/National Environmental Satellite, Data, and Information Service (NESDIS), and the U.S. Geological Survey (USGS) to conduct a decadal survey to generate consensus recommendations for the Earth and environmental science and applications communities regarding a systems approach to space-based and ancillary observations that encompasses the research programs of NASA; the related operational programs of NOAA; and associated programs such as Landsat—a joint initiative of USGS and NASA.

The last Decadal Survey for Earth Science and Applications from Space was published in 2007 (National Academy of Sciences 2007). The Decadal Survey for 2017–2027 is now ongoing. In the current survey, its "prioritization of research activities will be based on the committee's consideration of identified science priorities; broad national operational observation priorities as identified in U.S. government policy, law, and international agreements (for example, the 2014 National Plan for Civil Earth Observations) and the relevant appropriation and authorization acts governing NASA, NOAA, and USGS; cost and technical readiness; the likely emergence of new technologies; the role of supporting activities such as in situ measurements; computational infrastructure for modeling, data assimilation, and data management; and opportunities to leverage related activities including consideration of interagency cooperation and international collaboration…The survey committee will work with NASA, NOAA, and USGS to understand agency expectations of future budget allocations and design its recommendations based on budget scenarios relative to those expectations. The committee may also consider scenarios that account for higher or lower than anticipated allocations…The committee may also identify potential interagency and international synergies; proposed augmentations to planned international missions; and adjustments to U.S. missions planned, but not yet implemented."

The use of Earth observations for policy include a variety of areas such as: agricultural crop production, atmosphere–ocean coupled general circulation models (with CO_2 damages being quantified and the dollar value of damage being applied to the cost of new appliances for purposes of establishing minimum energy efficiency standards in the U.S.), direct input to the work of the Intergovernmental Panel on Climate Change (IPCC), the application of Gravity Recovery and Climate Experiment-Data Assimilation System (GRACE-DAS) to the U.S. Drought Monitor Weekly Mapping Process, weather projections and forecasts; near real-time adaptive management or identification of deviation in long-term trends; and cyber infrastructure.

The aforementioned assessments and surveys rely upon various methods to judge impact and rank priorities. Earth observation projects almost always provide information as their primary output. If decision-makers who are the primary users of the project's output are available to assist, then Value of Information (VOI) methods can be used (NASA Earth Science Applied Sciences Program 2012).

3.2 Quantifying the Economic Value of Information

VOI methods, which are primarily prospective, measure how new information changes a decision-maker's prior beliefs about uncertainties and the value the decision-maker would derive from the resulting change (NASA Earth Science Applied Sciences Program 2012).

VOI is a concept that enables the integration of the economic value of satellite data into economic analyses. The idea that information has value in both a statistical and a pragmatic sense dates back at least to the 1950s. In recent years, interest in the economic VOI has taken center stage. In the field of space studies, this refers to innovations in the technologies that collect information; this is the new information provided by the growing number of Earth-orbiting satellites, an area where recent applications of VOI methods are critically important for informing investment in satellite networks (Laxminarayan and Macauley 2012).

The following are VOI principles that are relevant for Earth observation:

- Information has value if it can either make a current choice more secure and confident or if it can reveal a different choice as better than the current choice;
- Information may lead to recognition that there is more uncertainty than initially thought;
- Information has value even if it introduces more uncertainty (it reveals that what was thought to be certain may not be);
- Perfect information may not be worth the cost of acquisition;
- Some attributes of information may confer more value than others (e.g., spatial, temporal, and spectral resolution; accuracy, precision, and other statistical properties of the distribution of information);
- Information tends to have value if it enables an action or a decision;
- Information, once acquired, is available at low additional cost, and one person's use may not preclude another person's use. Some information is about goods and services for which no prices exist.

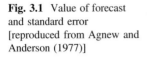 **Fig. 3.1** Value of forecast and standard error [reproduced from Agnew and Anderson (1977)]

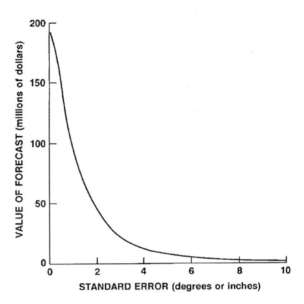

3.2.1 Forecast Error Contribution and VOI

As early as 1977, Agnew and Anderson described the exponential relationship between standard error and the value of forecast (see Fig. 3.1).

Forecast Error Contribution (FEC) can be used to measure the impact of individual observing systems, and therefore the value of their information. Figure 3.2 shows, for 24 h forecasts, the distribution of FEC (in percent) for the different observation types summed (top panel) and single observation normalized (bottom panel). It shows that the Advanced Microwave Sounding Unit-A (AMSU-A) is the single most valuable source of data, while buoy data are one of the most important components of the conventional observing system. According to the bottom panel, buoy impact per single observation is by far the largest among all different kinds of observations (Radnóti et al. 2012).

3.2.2 Examples and Further Reading

There are various methods and examples of Earth observation VOI analyses. Bernknopf et al. (2012) describe an economic model that involves application of geospatial data to land use and air quality regulation. Bouma et al. (2009) conducted a study on monitoring water quality. Econometric modeling and estimation examples include studies on agricultural productivity (Tenkorang and Lowenberg-DeBoer 2008); life expectancy (Obersteiner et al. 2012); and other quality of life dimensions. Simulation modeling and estimation is another method,

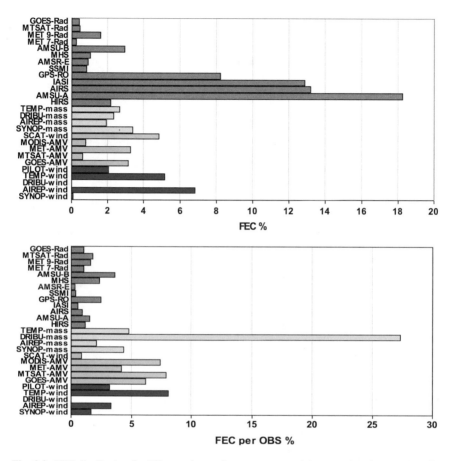

Fig. 3.2 FEC distribution for different observation types summed (*top panel*) and normalized for a single observation (*bottom panel*). The FEC is displayed in % and it is derived for 24 h forecast errors. Reproduced from Radnóti et al. (2012)

while price- and cost-based derivation studies include weather data for weather insurance (Osgood and Shirley 2012); drought and land use information for index insurance (Skees 2008); and losses averted from vector-borne disease (Hartley 2012). Probabilistic approaches include Bayesian belief networks (Kousky and Cooke 2012) and expert elicitation (Interagency Working Group on Social Cost of Carbon 2010). These studies are described in detail in "The Value of Information" (Laxminarayan and Macauley 2012).

References

Agnew CE, Anderson RJ (1977) The economic benefits of improved climate forecasting. Princeton, New Jersey

Bernknopf RL et al (2012) Estimating the benefits of land imagery in environmental applications: a case study in nonpoint source pollution of groundwater. In: Laxminarayan R, Macauley MK (eds) The value of information: methodological frontiers and new applications in environment and health. Springer, Netherlands, pp 257–299

Bouma JA et al (2009) Assessing the value of information for water quality management in the North Sea. J Env Man 90(2):1280–1288

Executive Office of the President (2010) National space policy of the United States of America. https://www.whitehouse.gov/sites/default/files/national_space_policy_6-28-10.pdf. Accessed 29 Nov 2016

Future of Land Imaging Interagency Working Group, Executive Office of the President (2007) A plan for a U.S. national land imaging program. https://eros.usgs.gov/sites/all/files/external/eros/history/2000s/2007%20A%20Plan%20for%20U.S.%20National%20Land%20Imaging%20Program.pdf. Accessed 29 Nov 2016

Hartley DM (2012) Space imaging and prevention of infectious disease: rift valley fever. In: Laxminarayan R, Macauley MK (eds) The value of information: methodological frontiers and new applications in environment and health. Springer, Netherlands, pp 231–243

Interagency Working Group on Social Cost of Carbon (2010) Social cost of carbon for regulatory impact analysis. https://www.whitehouse.gov/sites/default/files/omb/inforeg/for-agencies/Social-Cost-of-Carbon-for-RIA.pdf. Accessed 29 Nov 2016

Kousky C, Cooke RM (2012) The value of information in a risk management approach to climate change. In: Laxminarayan R, Macauley MK (eds) The value of information: methodological frontiers and new applications in environment and health. Springer, Netherlands, pp 19–43

Laxminarayan R, Macauley MK (2012) The value of information: methodological frontiers and new applications in environment and health. Springer, Netherlands

NASA (2012) Applied sciences program 2011 annual report. Washington, D.C., USA

National Academy of Sciences (2007) Earth science and applications from space: national imperatives for the next decade and beyond. https://www.nap.edu/catalog/11820/earth-science-and-applications-from-space-national-imperatives-for-the. Accessed 29 Nov 2016

National Science and Technology Council, Executive Office of the President (2013) National strategy for civil Earth observations. https://www.whitehouse.gov/sites/default/files/microsites/ostp/nstc_2013_earthobsstrategy.pdf. Accessed 29 Nov 2016

National Science and Technology Council, Executive Office of the President (2014) National plan for civil earth observations. https://www.whitehouse.gov/sites/default/files/microsites/ostp/NSTC/national_plan_for_civil_earth_observations_-_july_2014.pdf. Accessed 29 Nov 2016

Obersteiner M et al (2012) Valuing the potential impacts of GEOSS: a systems dynamics approach. In: Laxminarayan R, Macauley MK (eds) The value of information: methodological frontiers and new applications in environment and health. Springer, Netherlands, pp 67–90

Osgood D, Shirley KE (2012) The value of information in index insurance for farmers in Africa. In: Laxminarayan R, Macauley MK (eds) The value of information: methodological frontiers and new applications in environment and health. Springer, Netherlands, pp 1–18

Radnóti G et al (2012) ECMWF study to quantify the interaction between terrestrial and space-based observing systems on numerical weather prediction skill. http://www.ecmwf.int/sites/default/files/elibrary/2012/11814-ecmwf-study-quantify-interaction-between-terrestrial-and-space-based-observing-systems.pdf. Accessed 29 Nov 2016

Skees JR (2008) Innovations in index insurance for the poor in lower income countries. Agric Resour Econ Rev 37(1):1–15

Tenkorang F, Lowenberg-DeBoer J (2008) On-farm profitability of remote sensing in agriculture. J Terre Obs 1(1):50–59

Author Biographies

Masami Onoda is currently the U.S. and multilateral relations interface at the International Relations and Research Department of the Japan Aerospace Exploration Agency (JAXA). As an academic, she is fellow of the Institute of Global Environmental Strategies (IGES) and she is also engaged in the private sector as an advisor to the Singapore-based space debris start-up Astroscale Pte. Ltd. since its foundation in 2013. From 2009 to 2012, Dr. Onoda was a scientific and technical officer at the intergovernmental Group on Earth Observations (GEO) Secretariat in Geneva, Switzerland. From 2003 to 2008, while pursuing her graduate studies, she was invited to the JAXA Kansai Satellite Office in Higashiosaka as a space technology coordinator to support technology transfer to SMEs for the small satellite project SOHLA-1. From 1999 to 2003, she worked in the field of Earth observations at JAXA (then NASDA), serving on the Secretariat of the Committee on Earth Observation Satellites (CEOS). In 1999, she was seconded to the UN Office for Outer Space Affairs (UNOOSA) for the organization of the UNISPACE III conference. She holds a Ph.D. in global environmental studies (2009) and a master's degree in environmental management (2005), both from the Kyoto University Graduate School of Global Environmental Studies. Her undergraduate degree is in international relations from The University of Tokyo.

Molly Macauley was vice president for research and senior fellow with Resources for the Future (RFF), Washington, DC. RFF was established at the request of President Truman as a think tank focusing on the economics of natural resources. Her research emphasized new technology and its application to natural and environmental resources, including the value of satellite-derived Earth science information and its use in understanding ecological systems and human relationships with these systems, and the value placed by the public on the nation's space activities. Dr. Macauley frequently testified before Congress and served on national-level committees and panels, including the Committee on Earth Science and Applications from Space of the National Research Council's Space Studies Board, the Board of Advisors for the Thomas Jefferson Public Policy Program at the College of William and Mary, the Science Advisory Board of the National Oceanic and Space Administration, the Earth Science Applications Advisory Group of NASA, and the Scholarship Committee of the Women in Aerospace Foundation. She also served as a lead author for NASA on a synthesis and assessment report for the U.S. Climate Change Science Program on the use of Earth observations. She was selected as a "Rising Star" by the National Space Society, was elected to membership of the International Academy of Astronautics, served as a distinguished visitor at the invitation of the Government of Quebec, and received awards for her work from NASA and the Federal Aviation Administration. Molly published extensively with more than 80 journal articles, books, and chapters of books. She was also a visiting professor in the Department of Economics at Johns Hopkins University. Her Ph.D. and M.A. degrees in economics were from Johns Hopkins University and her undergraduate degree in economics was from the College of William and Mary.

In a tragic incident Dr. Macauley passed away on July 8, 2016. This book is dedicated to Dr. Macauley, an inspirational contributor to the project, whose expertise in space policy and economics will always provide invaluable guidance to the space community.

Chapter 4
Benefits Assessment of Applied Earth Science

Lawrence Friedl

4.1 Introduction

Fresh water, air quality, deforestation, food security, urbanization, sanitation, land management, disease, biodiversity, hygiene, economic growth, and disasters. These and many others are all global challenges with environmental and resource dimensions. Increasingly, people and organizations are using Earth observations and scientific information about the Earth to gain insights on and inform their policy and management decisions related to these challenges.

Along with numerous organizations globally, NASA has been a key contributor to the wealth of data and information about the Earth and its processes. In addition, NASA has helped advance global knowledge about effective ways to apply the data and information across sectors and thematic areas. There are countless examples on how organizations have and are using Earth observations to support specific analyses, decisions, and associated actions.

Many of these examples have been qualitative and anecdotal. The substantiation of Earth observation benefits in societal and economic terms poses key challenges, yet this quantitative substantiation is of strategic importance to the Earth observation community as it expands efforts to inform decisions. It is at the heart of the value proposition. In addition, it comes at a time when there are increasing efforts to encourage greater integration of the social and economic sciences with natural sciences as well as global efforts to use data and indicators to address sustainability.

This chapter describes NASA's work to enable uses and applications of Earth observations, as well as efforts to quantify the socioeconomic impact and showcase the overall value of space-based observations.

L. Friedl (✉)
NASA Applied Sciences Program, Earth Science Division, Science
Mission Directorate, NASA Headquarters, Washington, D.C. 20546, USA
e-mail: LFriedl@nasa.gov

© The Author(s) 2017
M. Onoda and O.R. Young (eds.), *Satellite Earth Observations and Their Impact on Society and Policy*, DOI 10.1007/978-981-10-3713-9_4

4.2   Earth Science and Applications

One of NASA's agency-wide goals is to "advance understanding of Earth and develop technologies to improve the quality of life on our home planet" (NASA 2014). NASA's Earth Science Program supports fundamental and applied research on the Earth system, discovering new insights about the planet and the complex interactions within the Earth system.

Using the vantage point from space, global perspectives enable NASA to provide a broad, integrated set of high-quality data covering all parts of the planet. NASA shares this unique information openly with the global community, including members of the science, government, industry, education, and policy-maker communities (NASA 2014).

Within NASA Earth Science, its Applied Sciences Program is a dedicated effort to promote innovative and practical uses of Earth observations. The Program supports applied research and applications projects to enable near-term uses of Earth observations that inform organizations' decisions and that build key capabilities in the Earth science community and broader global workforce. Projects are carried out in partnership with private and public-sector organizations to achieve sustained uses and benefits from Earth observations. The Program addresses technical and human barriers to the use of new data and tools, creating new knowledge about effective methods and processes for applying Earth science (e.g., Hossain 2014).

4.3   Inform Decisions

Within the scientific community, the identification of the human role in and impacts on the Earth system have grown significantly. There are clear human dimensions to physical and biological parameters, such as air quality conditions and land use patterns.

With this recognition have come numerous efforts supporting greater integration of natural and social sciences. A U.S. National Academy of Sciences report suggested efforts "to facilitate crosscutting research focused on understanding the interaction among the climate, human, and environmental systems ..." (NRC 2009). The Future Earth initiative stemmed from the proposal for "a new contract between science and society in recognition that science must inform policy to make more wise and timely decisions ..." (ICSU 2012). In the U.S., the Obama Administration directed U.S. Federal agencies in 2015 to factor the value of ecosystem services into Federal planning and decision-making (OMB 2015). The U.S. Global Change Research Program's (USGCRP) 2012 Strategic Plan also included an objective on greater integration with social, behavioral, and economic sciences (USGCRP 2012).

This integration also offers significant opportunities to improve decision-making. Knowledge from the decision sciences, economics, and other social sciences can support ways to better incorporate Earth observations into analyses used for policy and business management decisions. Notably, the USGCRP's Strategic Plan introduced a goal focused on informing decisions for the first time in the USGCRP's history. The language of policy and business management is often an economics-based lexicon. The integration will require adjustments within the natural science and Earth observation communities (Mooney 2013). However, the integration can help the Earth observation community gain skills to articulate its value and improve its support to decision-making.

It is with these trends that NASA has supported significant efforts to quantify the socioeconomic benefits of Earth observation applications and build familiarity within the Earth observation community.

4.4 Socioeconomic Benefits of Earth Observations

As suggested, national and international organizations are placing greater emphasis on the benefits achievable from applications of Earth observations. The determination of specific societal and economic impacts, especially quantitatively, can be challenging, yet these determinations are critical to the value proposition of Earth observations and to induce greater use.

NASA Earth Science and its Applied Sciences Program have supported numerous studies to assess and document the benefits of Earth observations for decision-making. NASA has and continues to advance analytic techniques and quantitative methodologies for determining socioeconomic impacts across a range of themes.

Some of the socioeconomic studies that the Applied Sciences Program has sponsored include

- Disasters: Volcanic ash and aviation safety;
- Water: Improving water quality management;
- Health: Malaria early warning using Earth observations;
- Drought: Value of information for the U.S. Drought Monitor;
- Ecosystems: Fisheries management and pelagic habitats;
- Air Quality: Enhancements to the BlueSky emissions assessment system;
- Wildfires: Benefits of BlueSky for smoke management and air quality; and,
- Air Quality: Earth observations and the Environmental Protection Agency's (EPA) AIRNow system.

The following are summaries of two project-impact studies that the Applied Sciences Program has sponsored; the information is paraphrased from the authors' reports and also appeared on the program's website.

4.4.1   Earth Observations and Air Quality

Air pollutants can cause significant short-term- and long-term effects to human health. The U.S. EPA operates the AIRNow air quality system, which health officials use to alert the public about hazardous pollution.

A NASA-sponsored project pursued the use of Aura, Aqua, and Terra data within the EPA AIRNow air quality system. By incorporating the Earth observations into AIRNow, EPA could expand the system's coverage to reach millions of people not currently covered by the network of ground-based air quality monitors.

In the project's economic impact report, the analysis involved two approaches: face-to-face interviews in three case study locations (Denver, Colorado; Atlanta, Georgia; and Kansas City, Missouri) to assess the public value or community-level benefits and analysis of cost savings from the use of satellite data instead of installing new monitors to provide air quality information for public health decisions to populations in currently unmonitored locations.

The study found that the addition of satellite data could provide daily particulate matter information to 82% of people living in currently unmonitored locations (approximately 15 million people); the study estimated that the capability represents a value of about USD 26 million.

The three case studies also identified nonmonetary value and benefits. Interviewees reported reduced adverse health impacts on sensitive populations resulting from more accurate air pollution warnings and health alerts, and increased public viewing and understanding of air quality maps on AIRNow because of greatly increased spatial coverage. They also reported increased media use of AIRNow air quality maps resulting from expanded geographic coverage; more comprehensive air quality stories available to the media because of improved geographical representation of pollutant transport resulting from unusual events; and better communication with the public about the spatial distribution of air pollution, especially in sparsely monitored areas, resulting in better public understanding of these issues.

4.4.2   Volcanic Ash, Earth Observations, and Aviation Safety

Large volcanic eruptions can eject ash to heights at which commercial aircraft normally fly. Volcanic ash can cause damage to engines and fuselages, making it necessary to reroute, delay, or cancel flights to protect aircraft and ensure passenger safety. The international aviation community uses information and warnings from nine Volcanic Ash Advisory Centers (VAAC) on the location of volcanic ash.

In 2010, Iceland's Eyjafjallajökull volcano erupted, sending volcanic ash into European airspace and canceling flights. European VAACs had not used Aura data, and a NASA-sponsored project team developed and delivered data products within days of the eruption. European officials used the Aura products in their determinations of which airspace to open.

An impact analysis analyzed the benefits of the project and VAAC's use of Aura data. One part focused on the benefits from use following the Eyjafjallajökull eruption, and one part focused on a global estimate of average annual benefits.

The analysis team used data on flight cancelations and revenue losses due to Eyjafjallajökull, historical frequencies of aircraft damage from volcanic ash, and aircraft repair costs. The team estimated how much the Aura data would reduce the uncertainty about the level of ash threat, determining a risk-adjusted value of the observations. Overall, the analysis found that the satellite data reduced the probability of an aircraft experiencing a volcanic ash incident by approximately 12%.

The team estimated that use of the data following the Eyjafjallajökull eruption saved USD 25–72 million in avoided revenue losses due to unnecessary delays and avoided aircraft damage costs. If the data had been used from the beginning of the incident, an estimated additional USD 132 million in losses and costs might have been avoided.

The team extrapolated the risk-adjusted results globally to estimate the potential annual impact from the use of Earth observations by VAACs. Accounting for annual frequency and magnitude of volcanic eruptions, the team estimated an expected value of up to USD 10 million annually.

4.5 Sustainable Development Goals

These efforts to measure the impacts of Earth observations on decision-making mirrors and aligns with other endeavors globally to use data to advance social, economic, and environmental progress. Most notable are the United Nations' Sustainable Development Goals.

In 2015, the United Nations endorsed Transforming Our World: the 2030 Agenda for Sustainable Development, a global development agenda for all countries and stakeholders to use as a blueprint for progress on sustainability. The 2030 Agenda specifically calls for new data acquisition and exploitation of a wide range of data sources to support implementation, including a specific reference to Earth observations and geospatial information.

Thus, the 2030 Agenda represents a key opportunity for Earth observations to play insightful roles in monitoring targets, planning, and tracking progress that can contribute toward achieving the goals. The long-term collection of data on the goals provides worldwide opportunities to produce examples and further develop analytic methods on the socioeconomic benefits of Earth observations.

4.6 Conclusion

Earth observations are often one of many sources of information that are used in decision-making activities. The articulation of the specific impact from Earth observations on a decision can be challenging to substantiate fully, yet the pursuit has both intellectual and strategic benefits. Similarly, a familiarity with the terms and methods of the social and economic sciences provide the Earth observation community with greater opportunities and audiences.

The global community has made significant strides in the past decade to begin tapping satellite data for the purposes of policy and management decisions. Thus, further efforts to determine and document Earth observations' impacts in socially and economically meaningful terms can support broader efforts to employ the data, seek additional data, and lead to additional advances and further societal benefits.

References

Hossain F et al (2014) A guide for crossing the valley of death: lessons learned from making a satellite based flood forecasting system operational and independently owned by a stakeholder agency. http://journals.ametsoc.org/doi/pdf/10.1175/BAMS-D-13-00176.1. Accessed 7 Dec 2016
ICSU (2012) State of the planet declaration. http://www.icsu.org/rio20/home/pup-state-of-planet-declaration. Accessed 7 Dec 2016
Mooney HA et al (2013) Evolution of natural and social science interactions in global research programs. Proc Natl Acad Sci USA 110:3665–3672
NASA (2014) 2014 Science Plan. Washington, D.C., USA
NRC (2009) Restructuring federal climate research to meet the challenges of climate change. National Academies Press, Washington, D.C.
U.S. Global Change Research Program (2012) The national global change research plan 2012–2021: a strategic plan for the U.S. Global Change Research Program. https://downloads.globalchange.gov/strategic-plan/2012/usgcrp-strategic-plan-2012.pdf. Accessed 7 Dec 2016
U.S. Office of Management and Budget (2015) Incorporating ecosystem services into federal decision making. https://www.whitehouse.gov/sites/default/files/omb/memoranda/2016/m-16-01.pdf. Accessed 7 Dec 2016

Author Biography

Lawrence Friedl serves as the director of the Applied Sciences Program within the Earth Science Division at NASA Headquarters. The program supports efforts to discover and demonstrate innovative and practical applications of Earth science by government, business, and other organizations. He has been with NASA since 2002 and he has served as the program manager for air quality applications along with several other applications themes. Among his responsibilities, Lawrence is a vice-chair of the interagency U.S. Group on Earth Observations (USGEO) and is the NASA principal for the interagency Civil Applications Committee. He represents the United States in the Group on Earth Observations (GEO), serving as a co-lead on a GEO initiative on sustainable development goals (SDGs). He chairs the International Committee for Remote Sensing of Environment and serves on the National Space Club's Award Committee for Innovative Uses of Earth Observation Satellite Data.

Prior to joining NASA, Lawrence worked at the U.S. Environmental Protection Agency (EPA), focusing on applications of geospatial data and technology. He also served as a space shuttle flight controller in NASA's Mission Control Center for 15 missions, including several Earth science missions. He joined the U.S. federal government as a presidential management intern. Lawrence received a master's degree in public policy from Harvard University's Kennedy School of Government, specializing in science and technology policy. He received a bachelor's degree in mechanical and aerospace engineering from Princeton University. He also received a certificate in space policy and law from the International Space University.

Chapter 5
ESA's Earth Observation Strategy and Copernicus

Josef Aschbacher

In 2015, ESA adopted a new *ESA Earth Observation Strategy 2040*. The former strategy had been written some 20 years ago and was closely linked to the concept of the Living Planet Programme. With the Earth Observation Envelope Programme (EOEP) and the Earthwatch mechanism, the strategy overcame the approach of running isolated, one-off missions. The advent of Copernicus, discussed in more detail below, was a key factor in the decision to prepare an updated strategy. *ESA's Earth Observation Strategy 2040* covers the period from 2015 to 2040, with an interim progress review scheduled for 2025.

The new strategy builds upon existing elements, but also takes into account new developments and boundary conditions. It responds to societal challenges such as food, water, energy, climate, and civil security, while fully embracing the novel opportunities of the ICT revolution. Cloud-based access to Earth observation data, for the benefit of all levels of society, is a central element.

According to the new Strategy

> The vision of ESA is to enable the maximum benefit of Earth observation for science, society and economic growth in Europe, served by European industry. ESA will implement this vision through its Earth observation programmes, working in close cooperation with Member States, the EU, EUMETSAT and European industry within the widest international framework.

The Strategy sees ESA in an enabling, facilitating, and leading role, taking account of the strategies, objectives, and activities of the other Earth observation stakeholders in Europe. It also recognizes that: "*economic factors are of increasing importance in decision-making*" and attaches great importance to the service industry.

The *ESA Earth Observation Strategy 2040* is implemented through a number of programmes that are dedicated to both scientific (EOEP, Earthwatch, Climate Change Initiative) and operational demands (Copernicus, MetOp, Meteosat).

J. Aschbacher (✉)
Earth Observation Programmes, European Space Agency, Via Galileo Galilei,
00044 Frascati (RM), Italy
e-mail: josef.aschbacher@esa.int

© The Author(s) 2017
M. Onoda and O.R. Young (eds.), *Satellite Earth Observations and Their Impact on Society and Policy*, DOI 10.1007/978-981-10-3713-9_5

Copernicus is a case in point, illustrating many of the concepts mentioned above. Copernicus, previously known as the Global Monitoring for Environment and Security (GMES) programme, is the most ambitious Earth observation programme to date. It will provide accurate, timely and easily accessible information to improve the management of the environment, understand and mitigate the effects of climate change, and ensure civil security. In essence, Copernicus will help to shape the future of our planet for the benefit of all.

Copernicus consists of a complex set of systems that collect vast amounts of data from multiple sources, including Earth observation satellites and in situ sensors (e.g., ground stations, airborne/seaborne instruments) and provides users with reliable and up-to-date information through a set of services related to environmental and security issues. These services fall into six main categories: land management, the marine environment, atmosphere, emergency response, security, and climate change. The European Commission (EC), acting on behalf of the European Union (EU), is responsible for the overall initiative, setting requirements and managing the services.

The Copernicus Space Component features a new family of dedicated satellites, called Sentinels, specifically for the operational needs of the Copernicus program. The Sentinels provide a unique set of observations, starting with the all-weather, day and night radar images from Sentinel-1A, launched in April 2014. Sentinel-1B followed in April 2016. Sentinel-2A and Sentinel-2B, launched in June 2015 and in March 2017, deliver high-resolution optical images for land services. Sentinel-3A, launched in February 2016, provides data for services relevant to the ocean and land. Sentinel-4 and -5 will provide data for atmospheric composition monitoring from geostationary and polar orbits respectively. Sentinel-6 will carry a radar altimeter to measure global sea-surface height, primarily for operational oceanography and climate studies. In addition, the Sentinel-5 Precursor mission will reduce data gaps between ENVISAT (in particular the SCIAMACHY instrument) and the launch of Sentinel-5.

The Copernicus Space Component is managed by ESA and serves users' satellite data available through the Sentinels and other Copernicus Contributing Missions at national, European, and international levels. As the architect of the Space Component, ESA coordinates satellite data delivery. The ground segment, facilitating access to satellite data, completes the Copernicus Space Component.

Policy makers and public authorities are the main users of Copernicus. Information supplied by Copernicus is used to develop environmental legislation and policies or to make critical decisions in the event of an emergency, such as a natural disaster or humanitarian crisis.

Copernicus is a prime example of successful cooperation between the EU and ESA in space matters and it marks the transition from scientific to operational Earth observation for the environment and civil security. It is a quantum leap in terms of coverage, data volume, revisit frequency, and long-term availability of satellite data. The full integration of operational users into the information chain is at the heart of Copernicus.

Copernicus is also a model for the future evolution of the EU-ESA relationship, and there are important synergies. The R&D and space infrastructure management expertise of ESA can help fulfill EU policy requirements. The EC can integrate space and its applications into relevant sectorial policies and deliver the necessary funding.

The socioeconomic impact of Earth observation has become a key argument to justify European investments. For Copernicus, independent studies conducted by Price Waterhouse Coopers and Booz & Allen have shown that, on average, EUR 1 invested in Copernicus leads to an economic benefit of up to EUR 10 due to better decisions, more efficient policy implementation, as well as savings due to better preparedness in case of natural disasters (ESPI 2011). Former Vice-President of the EC, Antonio Tajani, estimated that Copernicus would create up to 83,000 jobs in the EU by 2030. As another example, the European Association of Remote Sensing Companies (EARSC) estimates the economic value generated by the use of satellite imagery in supporting winter navigation in the Baltic Sea to be between EUR 24 million and EUR 116 million per annum alone.

Copernicus illustrates the various economic potentials of space. Classical business opportunities arise across the whole spectrum of the value-added chain, from development and manufacturing (upstream) to operations and services (downstream). However, the commercial potential of space also exists at a higher level. By providing platforms and infrastructure for information exchange and informed decision-making, space applications contribute to better governance, more efficient use of resources, and increased competitiveness. Furthermore, space utilization can yield evidence for long-term processes that increase knowledge and understanding as a basis for responsible action. This aspect of space is especially relevant to tackling the challenge of climate change. Reducing the impact of climate change will not only lead to economic savings, it will also allow countering societal challenges caused, for example, by droughts or food shortages, which are one of the prime causes of migration. Last but not least, space activities create benefits that are not always quantifiable in financial indicators, but that are tangible in providing better global information about the Earth's state and threats to its environment and people, thereby enhancing citizens' safety and quality of life.

There is another aspect of Copernicus relevant to the economic potential of Earth observation, namely the issue of data and information policy. In fact, the benefits mentioned above will only occur on a large scale if access to space-based data and information is full, free, and open, as in the case of Copernicus.

The issue of the data policy for Copernicus as a public endeavor touches upon the question of handling Public Sector Information (PSI). In principle, there are two ways of handling PSI that is meant for release—it can either be sold or given away for free. Data and information created by publicly owned infrastructure holds considerable value. However, this does not mean that it is economically wise to commercialize the data. The example of Landsat—a U.S. satellite program that had its data policy changed to free and open access in 2008—shows that the utilization of space-based information and its commercial exploitation increase significantly when it is supplied at no cost. In 2011, over 100 times more Landsat data was downloaded than in 2007.

A study by EARSC shows that the same effect can be expected for Copernicus (EARSC 2012). Free and open access to Sentinel data boosts economic activity and raises public tax income. These benefits, though time delayed, will outweigh the income that could have been generated by selling the data instead.

It should be recalled that raw data holds little commercial value. The value is created/increased by value-added products and services derived from the raw data —another motivation for providing the raw data for free to support the development of such products/services and to increase industrial competence within this sector. Young entrepreneurs need an open data policy to develop innovative businesses. Without an open data policy, returns will be much lower and barely recover investment costs.

To sum up these economic considerations, the Copernicus program's success depends on the widest possible availability and use of Copernicus data and information. Without free and open access and a long-term commitment to data availability, companies and users would not have invested in integrating the program into their services.

Beyond economic considerations, there are a number of other reasons why a free and open data policy has been adopted for Copernicus. Firstly, in general the EU itself advocates free and open access models to favor innovation, as demonstrated in its 2020 strategy for smart, sustainable, and inclusive growth, or in the PSI and Infrastructure for Spatial Information in the European Community (INSPIRE) directives, which aim to ensure the widest possible dissemination of environmental information held by public bodies. Moreover, the two directives share the objective of enhancing the transparency and availability of public data. In addition, Copernicus is the European contribution to Global Earth Observation System of Systems (GEOSS), which is based on the principle of free exchange of crucial satellite data and information. Furthermore, numerous (international) organizations, for example the World Climate Research Programme (WCRP) or the European Science Foundation (ESF), have voiced their strong advocacy of free and open data policies.

Looking outside Europe, the U.S., China, and Russia grant free and open access to satellite data. Since the provision of data to users takes place within an open world market, protection of European data would not have stopped others offering and sharing their data. Instead, it would have simply increased the dependency of scientists and value adders on non-European data. This would have been a strategic setback for Europe.

Last but not least, Member States, who invested significantly in establishing the space component, also advocated for a free and open data policy. Local governments and public bodies echoed the sentiment.

Accordingly, after some discussion, an overall Copernicus data and information policy based on the concept of free and open access was decided through the EC's Delegated Regulation (EU) No. 1159/2013, supplementing Regulation (EU) No. 911/2010. The overall Copernicus data and information policy enshrines the principles that had already been established for the Sentinel satellites. The key elements are that there are no restrictions, neither on the use (commercial and noncommercial) nor on users (be they European or non-European); that a free version of any dataset is available on the relevant dissemination platform; and that data and information are available worldwide, without limitation in time.

The FAO Forest Resources Assessments (FRA) of 2015 used Landsat and not Sentinel-2 data due to the uncertainty around the Copernicus data policy at the time of decision. Other similar programs would not have invested in either the ground segment or data assimilation strategies without guaranteed access.

There are market segments of space-based Earth observation where commercial models apply, for example in the data range of a resolution better than two meters (e.g., TerraSAR-X, COSMO-SkyMed, or Pleiades). The Sentinels, however, do not cover this market segment. If in the future a Sentinel (e.g., for security purposes) addresses such a market segment, the free and open data policy will have to be reviewed.

It should be stressed that a free and open data policy does not need to endanger commercial business models of Earth observation data providers. This is demonstrated by the case of Guyana. Despite having full terrain coverage from public data sources, Guyana also requires and purchases coverage from commercial sources to obtain higher resolution data for forest inventories. Sentinel data are very important for Guyana, but it continues to use complementary commercial data.

Copernicus is now a tangible reality with five Sentinels in orbit and more to follow soon. The volume of disseminated data is huge and rapidly growing, with more than 74,000 users registered on the Sentinel Scientific Data Hub and more than 24 Petabytes of data having been downloaded. The Copernicus services are in place or about to reach their full performance, with excellent user uptake. One can conclude that the right decisions have been made—Copernicus is up and running, it puts Europe at the forefront of operational Earth monitoring, and it delivers all kinds of benefits to citizens in Europe and around the world.

References

EARSC (2012) About GMES and data: geese and golden eggs. http://earsc.org/file_download/134/Open+Data+study+Final+report.pdf. Accessed 7 Dec 2016

ESPI (2011) The socio-economic benefits of GMES. http://www.espi.or.at/images/stories/dokumente/studies/ESP_Report_39.pdf. Accessed 7 Dec 2016

Author Biography

Josef Aschbacher began his professional career at the European Space Agency (ESA) in 1990 as a young graduate at the Centre for Earth Observation (ESRIN), Frascati, Italy. From 1991 to 1993, he was seconded as ESA representative to Southeast Asia at the Asian Institute of Technology (AIT) in Bangkok, Thailand. From 1994 to 2001, he worked at the European Commission's Joint Research Centre (EC/JRC) in Ispra, Italy, where he was, in his last position, the scientific assistant to the director of the Space Applications Institute. He returned to ESA in 2001 in the role of programme coordinator at ESA HQ, Paris, where he was responsible for advancing Copernicus activities within ESA and for international cooperation. He was promoted in 2006 to head of the Copernicus Space Office, where he led all activities for Copernicus within the Agency and with external partners, in particular the EC. In 2014, he was promoted to head of programme planning and coordination in the Directorate of Earth Observation Programmes, where he was responsible for planning the future actions and programs of the ESA Earth Observation Directorate and for formulating and implementing programmatic and strategic decisions across the Directorate. In 2016, he was appointed as ESA director of Earth observation.

Chapter 6
Earth Observation—UK Perspective and Policy

Martin Sweeting

6.1 Observations Providing Scientific Evidence to Support Decision-Making

Observations of the physical and built environment are of critical importance to every country. Since these environments are directly tied to national wellbeing, prosperity, and security, robust observing systems are vital for understanding, managing, and forecasting environmental change. The UK is no exception to this and it is important to capitalize on such observations to support decision-making in government with accurate and timely scientific evidence for the greatest public benefit.

Knowledge of the consequences of urban and rural development on the quality of life, as well as the efficiency of business, can only be gained through comprehensive, continuous, and fresh observational data derived from a wide variety of sensors and sources. Comprehensive environmental observation also requires systems that can be applied to various timescales, from short-lived events that need to be monitored on an hourly or daily basis to long-term climate and geological changes. Robust, continuous and accurate data are required to help the UK Government make evidence-based decisions about the effects of climate change and their possible mitigation. These observations need to be assured over the long-term so that they can provide information on trends such as the rate of the global rise in sea level, the contribution of greenhouse gases and the effect of air traffic on the solar radiation budget.

M. Sweeting (✉)
Surrey Satellite Technology Limited, 20 Stephenson Road, Surrey Research Park, Guildford GU2 7YE, UK
e-mail: M.Sweeting@SSTL.co.uk

© The Author(s) 2017
M. Onoda and O.R. Young (eds.), *Satellite Earth Observations and Their Impact on Society and Policy*, DOI 10.1007/978-981-10-3713-9_6

6.2   Challenges for UK Society and Role of Observations

The UK faces a number of specific challenges: a growing and ageing population necessitating urban development and improved infrastructure services; a shift from traditional industries to an information-based economy that requires ubiquitous communications; increased severity of impact and frequency of natural events; and a need for actionable knowledge to counter the threats posed by terrorism. The need to adapt to the changing environment necessitates the monitoring of a growing range of parameters that feed into increasingly sophisticated models, using information derived from observation data to provide a more comprehensive picture on which government departments can base strategic, policy, and tactical decisions. These decisions will have significant impact on the UK economy, as well as the health and safety of its citizens. There are, however, a number of challenges in capitalizing on these almost overwhelming streams of sensor data. Exponentially growing databases need to be stored, processed and transformed into concise information products that are useful for knowledge-based decision-making across government, industry, and business.

Significant variations in prevailing UK climate conditions are already affecting housing, agriculture and livestock, and the reliability of transport. The UK observation policies and planning aim to address both monitoring these near-term phenomena and long-term trends. An understanding of the impact upon the UK of climate changes occurring beyond our shores is, of course, necessary and relies upon observations that are global and long-term, with international standards of interoperability so that data can be shared and used globally.

Environmental observation from space is a fast-moving field. Sensor and instrument technologies are developing rapidly, along with the platforms to carry them, especially the recent deployment of constellations of smaller, lower cost satellites and/or networks of sensors alongside larger multi-instrument platforms. Combined with tremendous advances in storage and data-processing capabilities, there are unprecedented opportunities for space-borne instruments to enhance our understanding of the dynamic environment. However, to take best advantage of these advances the UK sees that there is an urgent need for an integrated, coordinated, and sustained international climate observing system.

6.3   New Technologies and Opportunities

There are new techniques emerging and on the horizon that will dramatically change the field of Earth observation, both in technological capability and the need for new policies and, possibly, legal frameworks.

Constellations comprising hundreds of highly capably small satellites (such as those proposed by Google, Facebook, SpaceX, UrtheCast, OneWeb), operating alongside conventional large-scale science observatory missions, will provide agile global monitoring with a variety of rapidly developed sensors on a greatly expanded scale and persistence.

Possibly the most significant change may come from automated sensors connected to the internet for real-time data as part of the 'internet of things'. This, coupled with the wider use of opportunistic data from systems not originally designed for environmental observation (hand-held devices, cars, aircraft, etc.), could result in an entirely new dimension to data gathering, creating both high spatial and temporal observations. The emergence of crowd sourcing and 'citizen science', when combined with authoritative data sources, could provide unprecedented feedback on the effectiveness of policy and timely evidence of environmental impact. These developments will cause an avalanche of data. Cloud computing techniques will be employed for the effective processing, storage, and dissemination of the observations and enable to powerful data mining and cross-correlation on a scale almost unimaginable hitherto. However, it is critical to move these new technologies from research to operational deployment in a timely manner.

The resulting information base will provide policy makers with unprecedented opportunities for extracting actionable knowledge to respond to near-real-time events, such as floods and pollution, as well as longer duration trends, such as tracking sea level rise and the shrinking of polar ice caps and forests. Planning for the future is based on understanding the past, the present, and trends. Forecasts and planning data are intended to provide consistent and timely information to governmental and commercial planners to reduce uncertainty in multiple sectors, e.g. transport, agriculture, insurance, energy, and healthcare.

6.4 The National and International Contexts

Potential policy and economic benefits of observing systems are tied to national interests and priorities, but some policy and legal aspects of environmental observing systems have regional or global implications and can only be addressed in an international context. Many international agreements and undertakings depend on high quality, fresh, geospatial, and environmental data. The UK is a relatively small island but, whilst it has its own specific concerns, it is inevitably affected by phenomena beyond its shores. For example, data with high spatial and temporal resolution on a regional basis are needed to assess air quality, attribute the source of air pollution, and identify pollution transport mechanisms to identify specific regional problems and to monitor for compliance with treaties on climate, stratospheric ozone depletion, etc.

The UK approach is to recognize that contributions by all forms and sources of observations, both national and international, are needed and that, if treated coherently, the value of this whole is greater than the sum of the parts. The UK therefore contributes substantially to international observing programmes (e.g. Copernicus) whilst also supporting national applications through the Harwell Applications Catapult to provide better and easier access to data and the assurance of data continuity that are essential for entrepreneurs and small to medium-sized enterprises (SMEs) to commit to developing information products and services. Thus, there is a growing realization and move towards dissolving data barriers and sharing public and private observational data between government departments, institutions, and commercial entities, thus encouraging a pan-government appreciation of the benefits of observations and cooperative exploitation for policy and action. With the right observation infrastructure, skills base, and international partnerships, the information gathered from these systems can drive social policy and economic growth.

6.5 The Importance of Sustained Funding and Stakeholder Dialogue

The UK government understands that sustaining services and research and making the necessary advances in monitoring and forecasting in the future requires stable long-term funding, accompanied by wise policies governing observation methods. The improved understanding of the natural environment, together with enhanced predictive capability in weather, climate and related hazards, create opportunities for industry and business in this developing global market.

There are, however, new policy challenges brought about by these advances. The increased ubiquity, coupled with spatial and temporal resolution, raises questions associated with individual privacy. Non-state rather than institutional players may soon dominate observation data banks and their exploitation for commercial benefit may give rise to conflicts of interest. Increasing ease and affordability of access to the space environment will bring into question present export policies, shutter control policies and drive regulations regarding orbital debris. The growth in capability of commercial UK small satellites for optical Earth observation and radar remote sensing has driven the UK Space Agency to develop a national Earth Observation policy for the first time.

To ensure that these critical issues are addressed and the best use of environmental observations is made for national interests, the UK understands that there is a clear need for continuing dialogue between government stakeholders, industry, business and the academic/expert community. This dialogue should address the need for effective, efficient exploitation of existing systems and evaluate opportunities for new capabilities for environmental monitoring. It is important to capitalize on environmental observations to support decision-making in government with accurate and timely scientific evidence for the greatest public benefit.

6.6   An Example of Novel International Collaboration on Earth Observation Led by the UK

In 2002, the UK formed and led the international Disaster Monitoring Constellation (DMC), a wide-area satellite surveillance system based around a constellation of small Earth observation satellites specifically to provide daily revisit anywhere in the world to assist in the monitoring, assessment, and mitigation of natural and man-made disasters.

The constellation comprised satellites contributed by different nations (Algeria, China, Nigeria, Spain, Turkey, UK), each built and launched by SSTL in the UK. A total of seven microsatellites and one minisatellite carrying 600 km wide-swath medium-resolution (~ 20 metres GSD) imagers have been launched into the same 686 km polar sun-synchronous orbit and manoeuvred so that they are equidistant around that orbit. This ensures that the wide imaging swaths provide the rapid revisit capability.

Each satellite is owned and operated in orbit by a different country and organization and used to meet their individual national Earth observation needs; however, the constellation has been coordinated by SSTL, through DMCii Ltd, to be able to provide rapid imagery in response to activations of the International Charter or directly to stricken areas of the constellation members. Members of the constellation also exchange data to be able to gain the benefits of a constellation of multiple satellites whilst only having to fund a single satellite. As disasters occur relatively infrequently, the excess capacity of the satellites was used extensively for monitoring agriculture, water and land resources, mapping, pollution, deforestation, and desertification, as well as providing data to commercial users. The 600 km wide imaging swath provided users with immediate large-area views, rather than having to be built up from a mosaic of scenes collected on different dates, sometimes weeks or months apart with a variety of lighting conditions. The wide swath is also a major advantage for disaster monitoring/relief operations; for example, the entire area of the Indian Ocean basin that was affected by the tsunami in 2004 was mapped by DMC satellites in less than 2 weeks, enabling comprehensive analysis of the worst-hit areas and the queuing of higher resolution (but limited swath) satellites from other Charter contributors to assess in greater detail.

For this reason, in addition to the wide-area payload, the fifth satellite in the series was equipped with a much higher resolution camera that permitted objects as small as 4 m on the ground to be distinguished and to provide more detailed follow-up observations of smaller areas (up to 24 km in extent), which were identified as being of particular interest through the analysis of the medium-resolution data. Subsequently, in 2011 and in 2015, more capable minisatellites with resolutions down to 1 m ground sample distance alongside improved medium-resolution wide-swath imagers were added to the DMC. The DMC proved to be a highly effective and novel form of partnership in Earth observation using small satellites.

Acknowledgements Chapter 3.4 was drawn from the report 'Observing the Earth: Expert views on environmental observation for the UK' that was commissioned by the UK Government Office for Science from The Royal Society and chaired by the author of this section. Ref: ISBN 9781782521471. https://royalsociety.org/~/media/policy/projects/environmental-observation/environmental-observations-report.pdf

Author Biography

Sir Martin Sweeting has pioneered rapid-response, low-cost, high-performance small satellite development since the foundation of Surrey Satellite Technology Ltd. (SSTL) as a university spin-off company 30 years ago. SSTL has grown today to a company of 550 employees and has built and launched 47 satellites including the European Galileo constellation and high-resolution Earth observation satellites. Prof. Sweeting is also distinguished professor at the Surrey Space Centre, University of Surrey, leading advanced space technology research. Prof. Sweeting was knighted by Her Majesty the Queen in 2002 for his services to small satellite engineering. He is fellow of the Royal Society and the Royal Academy of Engineering.

Chapter 7
Benefit Assessment of the Application of Satellite Earth Observation for Society and Policy: Assessing the Socioeconomic Impacts of the Development of Downstream Space-Based Earth Observation Applications

Murielle Lafaye

7.1 Introduction

Space provides a unique viewpoint for understanding our living planet and providing benefits for society. Satellite Earth observations can benefit many areas of society, including environment and resources management, agriculture and food security, transport, air quality and health, risk management, and security.

Since the start of the SPOT program, the French government has encouraged the use of satellite Earth observations for public policy needs and downstream applications. In response, the French National Centre for Space Studies (CNES) initiated an iterative dialog with policy makers, scientific communities, and industrial partners in order to support feasibility and demonstration studies as well as improve data accessibility, tools, and expertise.

Today, *NewSpace*[1] highlights the crucial role that space-based assets can play to meet population needs. It provides new opportunities for Earth observation to contribute to economic growth and jobs creation.

---

[1]*NewSpace* is a dynamic for the commercialization of low Earth orbit (LEO) satellites. Serving the ISS with commercial SpaceX or Blue Origin launches is a first step, which could be followed by the emergence of many new Earth observation service providers and massive Earth observation data flows.

M. Lafaye (✉)
Prospective Spatiale et Enjeux Socio-économiques, CNES – DIA/IP - Bpi 2903,
18 Avenue Edouard Belin, 31401 Toulouse Cedex 9, France
e-mail: murielle.lafaye@cnes.fr

© The Author(s) 2017   93
M. Onoda and O.R. Young (eds.), *Satellite Earth Observations and Their
Impact on Society and Policy*, DOI 10.1007/978-981-10-3713-9_7

In the context of financial constraints, the need to underline the return on investment (ROI) of government funding to the space sector has increased. Accounting for the return focuses on the applications and services downstream. In order to maximize return, CNES now considers ex ante identification of the socioeconomic benefits of Earth observation for public policy users and space-based service providers as a requirement for all projects engaged in the frame of the French Programme d'Investissement d'Avenir (PIA).

7.2 Assessing the Benefits of Satellite Earth Observation for Public Policies: A Focus on Environmental and Maritime Policies

Convincing policy makers to make increased and more effective use of space-based Earth observation data and products requires robust benefit and cost assessments that compare Earth observation solutions to existing frameworks. A clear understanding of the value chain and actors is critical. Quantitatively assessing the benefits of space-based Earth observation is challenging because different multi-discipline approaches are needed for each case.

Several other issues also have to be addressed in order to motivate policy makers to invest in space-based Earth observation: space asset continuity, data and product accessibility, reliability, performance, and affordability are all relevant concerns.

Facilitating access to space Earth observation data is a key issue. Efforts are ongoing to facilitate access to scientific data hubs, e.g., ForM@Ter,[2] AERIS (collating Ether and ICARE),[3] and Aviso.[4] Copernicus Sentinel data and processing facilities are also made accessible via Plateforme d'Exploitation des Produits Sentinel (PEPS).[5] SPOT World Heritage data are made accessible through Theia.[6]

Convinced that space Earth observation could be a promising tool for the implementation and control of public policies, the French Ministry of Environment (MEDDE) and Ministry of Overseas Territories (DGOM) have intensified their relationship with CNES to facilitate the integration of satellite Earth observation information and their decision-making systems.

---

[2]ForM@Ter: http://poleterresolide.fr/.

[3]AERIS: http://pole-ether.fr and http://www.icare.univ-lille1.fr.

[4]AVISO: https://aviso-data-center.cnes.fr/.

[5]PEPS: https://peps.cnes.fr.

[6]Theia: https://theia.cnes.fr/.

7.2.1 MEDDE PlanSat: Satellite Earth Observation Contribution Toward Environmental Policy

The relationship between CNES and MEDDE was initiated at the beginning of the SPOT program to raise awareness about the potential of Earth observation data and to facilitate its integration into environmental policy. Thanks to SPOT's multi-scale, long-term information series and high temporal resolution, rapid mapping, and change detection information can also be provided to risk management actors (civil security and health departments). Such value-added products can also be used to anticipate damages. It is expected that this Earth observation information can provide policy makers with major benefits.

Anticipating Copernicus data availability, MEDDE elaborated PlanSat[7] to identify space Earth observation contributions for directorates in charge of Prevention and Risks (DGPR), Energy and Climate (DGEC), Infrastructures-Transport and Maritime (DGITM), Town and Country Planning-Lodging and Nature (DGALN), Fishery and Aquaculture (DPMA) and Civil Aviation (DGAC). In 2011, MEDDE and CNES signed a master agreement to implement actions.

Quantifying the benefits of using space Earth observation within a domain is a significant undertaking. The method used by MEDDE and CNES to assess the benefits of the implementation and control of public policies was based on the concept of demonstration-comparison-feedback, with and without space Earth Observation assets. Assessments were undertaken for various monitoring and control issues in policy areas such as soil-use, infrastructure monitoring, coastal erosion, and natural hazard management.

Quantitative assessments of the benefit of space Earth observation necessitates the clarification of value chains in each domain and determination of the "value" of space-based Earth observation information within each segment of the chain. Input from space Earth observation experts, economists, social sciences experts, and policy makers is mandatory. CNES has also initiated discussions on space Earth observation value in the frame of a new group that gathers social sciences experts. Long-term sharing of common definitions and methods are conducted by CNES, MEDDE, and industrial partners within Le Comité de Concertation Etat Industrie sur l'Espace (COSPACE) in cooperation with Commissariat Général à l'Investissement (CGI)[8] and Direction Générale des Entreprises (DGE).[9]

---

[7]PlanSat is the MEDDE master plan that identifies space-based Earth observation benefits and needs for each directorate: http://www.developpement-durable.gouv.fr/Le-plan-d-applications.html.

[8]CGI is funding space technology transfer for industrial production capacities in the frame of PIA.

[9]DGE supports the development of the upstream and downstream space sectors.

7.2.2 DGOM: Satellite Earth Observation for Overseas Challenges

For territories overseas, France must implement and control public policies in various domains, including health, transportation, town and country planning, agriculture, natural risks and crisis management, environmental and maritime surveillance, illegal fishing, pollution surveillance, and security.

Needs are specific, depending on the geographical situation and on the geo-political challenges of the overseas territories located in the tropics, Arctic, or Antarctic. The Direction Générale des Outre-Mer (DGOM) is in charge of coordinating the efforts of the different ministries involved in the implementation and control of public policies overseas.

The benefit assessment of space Earth observation was conducted within the frame of a demonstration-comparison-feedback process. The relationship between DGOM and CNES was initiated in response to illegal fishing in French Guyana. CNES facilitated access to expertise at Collecte Localization Satellites (CLS) and supported a feasibility study for which satellite optical and radar sensors were combined to provide maritime security authorities with adapted products for information and decision-making. In 2013, convinced that satellite Earth observation was a promising tool, DGOM and CNES decided to cooperate on a long-term basis and a master agreement was signed.

Sargassum algae surveillance is one of the concrete actions for which a benefit assessment is ongoing. MEDDE, DGOM, and the Ministry of Health collaborated to establish an appropriate response to the health, environmental, and economic problems caused by the decomposition of algae around the beaches of Martinique and Guadeloupe. In the frame of their master agreement, DGOM asked CNES to evaluate the possibility and cost of implementing a space-based Sargassum algae surveillance system. A feasibility study was initiated with the objective of evaluating the ability of satellite Earth observation to detect Sargassum algae and to evaluate the suitability of Sentinel tasking for an operational service. The CLS facilities VIGISAT[10] and Mobydrift[11] were run using free satellite Earth observation data from Landsat-8 (optical) and Sentinel-1 (radar). Discussions were initiated with Direction de l'Environnement et de l'Aménagement du Littoral (DEAL) Martinique and Guadeloupe to understand the operational constraints and information needs, in order to provide them with adapted information products.

Thanks to this study, the dynamic of the Sargassum algae drift along Martinique and Guadeloupe was characterized. A service based on a satellite system combining optical and radar sensors was dimensioned and the running costs estimated. An iterative dialog with future users is ongoing in order to optimize service quality and costs.

---

[10]A maritime surveillance Geographic Information System (GIS) for visualizing information products.
[11]A drifting model.

The Chamber of Commerce of Guadeloupe estimates fishery and tourism economic losses related to algal blooms to be about EUR 5 million (USD 8 million) in the first quarter of 2015 alone (Valo 2016). The health impacts of SO_2 vapor coming from Sargassum algae decomposition is difficult to estimate, but will increase global damage and economic losses. These are key elements for assessing the benefits of using space-based Earth observation information in early warning information systems for Sargassum algae.

7.3 Assessing the Downstream Economy of Space-Based Earth Observation: Challenges, Methods and First Results

Growing demand for satellite Earth observation from policy makers contributes to the justification for the development of next-generation Earth observation satellites. Growing demand for value-added products will contribute to the development of downstream service providers. Policy makers and administrations will be able to procure adapted products and services, initiating a virtuous economic loop.

A healthy space-based Earth observation service provider ecosystem can be considered an indicator of the benefits of public investment in space. Its contribution to the global economy and employment is to be evaluated.

With increasing awareness of the potential economic benefits of space-based Earth observation and the downstream service provider industry, governments have been investing in business development programs, for example, the UK Catapult and French *Boosters* initiatives. However, assessing the impact of these investments is a challenge, as no common methodology has been identified by the space sector or economists. For this reason, CNES motivated DGE to join discussions in the frame of the OECD Space Forum. CNES has also initiated several actions to assess the French space-based Earth observation sector, evaluating financial value, and the impact of government investment toward downstream services.

7.3.1 Towards a Methodology for Assessing the Economy of the Downstream Space-Based Earth Observation Sector

The economy of the upstream space industry is surveyed regularly. Market reports are available addressing the launcher, satellite, and ground segments, but such reports do not exist for the downstream space-based Earth observation sector.

Determining any impact requires a clear view of the domain to be evaluated. So, the first question is what to count as part of the space Earth observation downstream sector.

This introduces some further questions:

1. What is the perimeter of the downstream space Earth observation sector? The delineation between upstream and downstream is not clear, nor accepted by the traditional actors. Moreover, new players coming from the ICT sector do not present themselves as space Earth observation service providers but just as service providers.
2. What is the service taxonomy to be used? There is no common classification for space-based Earth observation services. Reading through catalogs of space-based Earth observation service promoters and/or providers shows that taxonomies of service vary depending on the community addressed (scientific, agencies, industry, etc.).
3. For a given service, how should the importance of the space Earth observation data contribution be evaluated alongside other sources of information? Service offerings are increasingly a combination of many sources of information. Insider knowledge is often required to appreciate the exact contribution.
4. Is the service operational or not, and is it delivered on a commercial basis?
5. How should public organizations be considered when they could have commercial activities?

Discussions are ongoing in the frame of the OECD Space Forum, which contributes to common definitions. Segmentation into three space segments was proposed, setting up perimeters related to space activities, products, and services (Fig. 7.1).

As a member of the OECD Space Forum, CNES has supported several studies in order to assess the impact of public investment on the development of the downstream space-based Earth observation sector. This work has been conducted in the frame of COSPACE, in coordination with Groupement des Industries Françaises Aéronautiques et Spatiales (GIFAS) and DGE.

The first issue was to establish awareness of the French downstream space-based Earth observation sector. A mapping of French organizations with activities that are either linked with, or directly using, space Earth observation was established (Magellium 2014). Following analysis, a segmentation of actors was proposed and French space-based Earth observation service providers were identified. Secondly, an estimate of the financial value for this sector in France was established.

This approach promotes raised awareness of the financial soundness of the sector and establishes a useful reference. Revisiting these studies regularly will make the assessment of the impact of public investment in the downstream space-based Earth observation sector possible.

Linking perimeters with activities / products / services

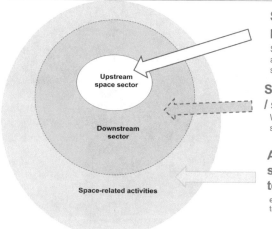

Space technology / product / service

Specific to the space sector – astronomy research, satellite, sub-system, …

Space-enabled product / service

Which WOULD not function without satellite capacity

Activities / products / services utilising space technology

e.g. ad-hoc spin-offs, technology transfers to non-space sectors…

Fig. 7.1 OECD Space Forum segmentation of the space sector

7.3.2 Mapping French Satellite-Based Earth Observation Service Providers

Within COSPACE, CNES has initiated discussions around elaborating require-ments for a study into mapping French space-based Earth observation service providers. A study, funded by CNES, was performed by Magellium in 2014 to test the following hypotheses:

- Societies with activities and jobs in France should be considered;
- A turnover of EUR 1 million is an effective threshold to divide up very small enterprises and industrial actors;
- Both commercial and noncommercial organizations should be identified;
- Segmentation of actors by sector of activity (software providers, data providers, service providers, etc.) is a requirement; and,
- "Real" service providers and their market segment should be precise.

CNES completed this study with an estimate of the financial value in 2015 (Fig. 7.2).

With a global picture of around 122 entities, downstream space-based Earth observation is well developed in France. The EUR 1 million threshold enables the identification of about 41 key actors with commercial activities in France. The 51

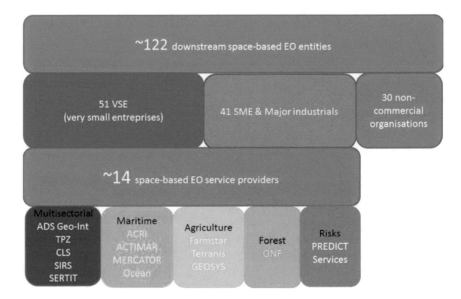

Fig. 7.2 Mapping downstream space-based Earth observation service providers in France (CNES 2015)

very small enterprises (one-expert, small consultancy offices, start-ups), spread France-wide, can be considered as a pool of resources that may grow with rising demand.

The study highlighted an ecosystem of about 14 providers serving policy makers and commercial users in traditional sectors such as agriculture, maritime, and disaster risks.

The landscape is expected to evolve considering the *Boosters* initiative and the arrival of new players from the digital economy, as well as the possible contributions of new systems such as UAVs and airborne Earth observation.

7.4   Lessons Learned and Perspectives

Just as governments are asked by taxpayers to provide the cost-effectiveness of their investments, space agencies must provide feedback on the benefits of using space and on the financial soundness of the ecosystems they develop. If the development of space assets to understand our living planet is, and remains, the main issue, providing taxpayers with an assessment of the benefits for public policies and the development of commercial space-based services is now part of the framework of Earth observation space programs.

Raising awareness of the capabilities of space-based Earth observation: assessing the benefit of a space-based Earth observation strategy begins with the utilization of space-based Earth observation data and products. Raising the awareness of public authorities and other potential Earth observation users remains a challenge. Promoting the capabilities of satellites is important, but not sufficient. Promoters of space-based Earth observation have to answer questions about service and cost benefits for users. Pertinent answers can only be given after an iterative cross-sensitization process that raises the awareness of both sides. Long-term cooperation in the frame of master agreements has proven its efficiency, providing both partners with a framework they can refer to in order to align actions, evaluate progress, and coordinate efforts.

Increasing the accessibility of space data: this is a key issue. Free and open data access will motivate users, especially, if the data are reliable and delivered continuously. Capacity building for potential users should be organized.

Assessing the value of space-based Earth observation data: studies should be conducted to identify the value of space-based Earth observation data. Supporting feasibility and demonstration studies with a user-driven approach is an important issue. Comparisons with existing tools/alternate methods should be organized in order to assess/quantify the benefits of using space-based Earth observation. This could be made possible by gathering policy makers and multidisciplinary experts from the Earth observation industry, social sciences, and economists. In support of the objective of providing stakeholders with quantitative information, long-term relationships should be organized.

Developing a common definition of space sector segmentation: the space manufacturing sector is well-structured and related economic surveys are conducted on a regular basis, but the same is not true for the downstream Earth observation sector. Work conducted within the OECD Space Forum will contribute to common definitions of space sector segmentation between upstream, downstream, and space-related segments. Comparisons between countries will then be possible.

Measuring the impact on the space-based Earth observation services ecosystem: involvement of policy makers in decisions related to the development of new satellites is a key indicator of the impact of space-based Earth observation services on their work. The Copernicus program is a major achievement of dialogue and partnership between policy makers, space agencies, scientific communities, and industrial partners. A public demand for geospatial information has been structured. The motivation of policy makers to use Sentinel data for public policies within the core services[12] is a key indicator and a trigger to develop an ecosystem of space-based Earth observation downstream services.

The French *Boosters* initiative aims to develop the use of space tools. By benefiting from the digital economy leverage, this initiative paves the way toward the development of flourishing new commercial services.

---

[12]Land monitoring, marine monitoring, atmosphere monitoring, emergency management, security, and climate change.

Regularly mapping the state of the art, CNES has organized itself to provide stakeholders with pertinent surveys of the downstream space-based Earth observation sector.

Providing opportunities for satellite Earth observation through the *NewSpace* context: new actors are entering the space sector as it has been identified as a promising economic domain. Coming from the digital economy, these actors are investing (or motivating venture capitalists to invest) in satellite Earth observation constellations, overthrowing the established order. Between 2014 and 2016, approximately 15 space Earth observation constellation projects were announced. Space agencies' free and open access data policies are facilitating the use of space Earth observation data for all. The challenge of using space-based Earth observation data has shifted from data accessibility to data dissemination and services. In these domains, new ICT players are leading the way. With their mastery of cloud-based hosting and big data analytics, new space actors are challenging traditional Earth observation service providers by democratizing the use of space and embedding space-based Earth observation information with other sources of information, making it difficult to evaluate the contribution and value of space-based Earth observation. As many services are turned into digital apps, the benefit assessment of space-based Earth observation will face new challenges.

Reference

Valo M (2016) Les algues sargasses, nouveau fléau des Antilles. http://www.lemonde.fr/planete/article/2016/02/19/les-algues-sargasses-nouveau-fleau-des-antilles_4868116_3244.html. Accessed 22 Aug 2016

Author Biography

 **Murielle Lafaye** is a foresight and economic impact expert at France's Centre National d'Études Spatiales (CNES) in the new Directorate of Innovation, Applications, and Science. Murielle has 18 years of experience in space-based downstream services development for various societal benefit areas (SBAs), such as health, agriculture, natural hazards and risk management, and overseas challenges. In her position at the Directorate of Strategy and Programs, Murielle was the point of contact with French ministries responsible for overseeing the overseas departments and territories of France, health, agriculture, and the digital economy. Her dual background in program management and business development involved her in the definition and implementation of the "Boosters" initiative to facilitate access to space data and accelerate space business within the frame of the digital economy. Fond of anticipating societal needs and space-based

services, Murielle is the coinventor of the CNES patent on tele-epidemiology. She has promoted the use of space for health in the frame of WHO, the Committee on Earth Observation Satellites (CEOS), and the Group on Earth Observations (GEO) as co-chair of the health SBA. In her new position, Murielle develops comprehensive scenarios to anticipate new trends for innovation and the space economy, in cooperation with policy makers, social scientists, digital economy stakeholders, and the space community in the frame of international organizations, the European Space Agency (ESA), and the OECD Space Forum.

Chapter 8
Chinese Earth Observation Program and Policy

Jinlong Fan

8.1 Development of the Chinese Earth Observation Program

With rapid economic development and global change, China is facing great challenges in managing natural resources, preserving the environment, and mitigating disaster risks. Chinese decision-makers have realized the important role that Earth observations have in effective and efficient management of Earth resources and reduction of disaster losses. In fact, Earth observation technologies have a special role in promoting sustainable economic and social development in all countries.

Before the year 2000, the major civil Earth observation satellite projects in China were limited to meteorological missions and the China–Brazil Earth Resources Satellite (CBERS). Investments were limited and developments were slow. In 2007, China (represented by 13 of its Ministries) released a 10-year plan for the China Integrated Earth Observation System at the Group on Earth Observations (GEO) Ministerial Summit in Cape Town, South Africa. The document outlined, for the first time, a prototype system of the Chinese Earth Observation Program. At almost the same time, the National Medium- and Long-term Plan for Science and Technology Development (2006–2020) proposed 11 key science and technology areas and 68 priority themes to be addressed in the next 15 years. Of these, 25 themes in eight key areas are directly related to Earth observations. In addition, the 11th National Five-Year Development Plan initiated 16 national key research projects, of which three are related to Earth observation science and technology. These developments highlight that Earth observation technologies are now recognized as major public goods that are urgently required in multiple areas.

J. Fan (✉)
National Satellite Meteorological Center, China Meteorological Administration,
Beijing, China
e-mail: fanjl@cma.gov.cn

© The Author(s) 2017
M. Onoda and O.R. Young (eds.), *Satellite Earth Observations and Their
Impact on Society and Policy*, DOI 10.1007/978-981-10-3713-9_8

A number of satellite series have been approved by the central government over the past 10 years. The Chinese Earth observation program uses a system-of-systems approach that incorporates meteorological, oceanic, and earth resources satellites, as well as private sector missions. The Integrated Chinese Earth Observations Program not only meets demands in China, but also makes contributions to the rest of the world. Chinese meteorological satellite data are distributed free of charge in line with international practice. International users may acquire data either through their own ground stations or the data portal operated by the National Satellite Meteorological Center (NSMC) of the China Meteorological Administration (CMA). At the 2007 GEO Ministerial Summit in Cape Town, South Africa, China also announced that CBERS satellite data were available free of charge. China has since built a CBERS ground station in South Africa (2008) and a HJ-1A ground station in Thailand (2011). China also promotes the use of high-resolution Chinese satellite data through the International Charter 'Space and Major Disasters.'

8.2   Key Components of the Chinese Earth Observation Program

8.2.1   Meteorological Satellites

The Chinese meteorological satellite program, FengYun (FY), was initiated in the 1960s and is composed of both polar-orbiting (odd numbered) and geostationary (even numbered) satellites. Each satellite is assigned an incremental alphabetic suffix according to its launch sequence.

The NSMC operates both the polar-orbiting and geostationary components, but the development and implementation of the FY program—from satellite development, launcher design, and manufacture, to telemetry, control, and ground applications—is the result of joint efforts between many institutes in China. The FY program is designed to enhance China's three-dimensional atmospheric sounding and global data acquisition capabilities, in an effort to collect more cloud and surface characteristics data from which meteorologists may infer atmospheric, land surface, and sea surface parameters that are global, all-weather, three-dimensional, quantitative, and multispectral.

FY-1 was China's first generation of sun-synchronous orbiting meteorological satellites. FY-1A and FY-1B were experimental satellites, launched on July 9, 1988 and September 3, 1990, respectively. FY-1C and FY-1D (operational satellites) were subsequently launched on May 10, 1999 and May 15, 2002. To date, all FY-1 satellites have completed their missions. The FY-1 series carried the visible and Infrared radiometer (VIRR) and space environment monitor (SEM). VIRR provided Earth imaging capability with ten channels at 1.1 km nadir spatial resolution, while SEM observed energetic particles in solar winds.

China started to develop its geostationary meteorological satellites, named FY-2, in the 1980s. The first satellite, the experimental FY-2A, was launched on June 10, 1997 and positioned at 105°E above the equator. This was followed by another experimental satellite, FY-2B. The subsequent FY-2C, FY-2D, and FY-2E satellites had noticeably enhanced performance; for example, the onboard scanning radiometer had five channels rather than the three of its predecessors. The latest satellites in the series, FY-2F, FY-2G, and FY-2H, are designed to serve as a consistent and stable bridge to the second generation of Chinese geostationary meteorological satellites, as the first generation fleet approaches the end of its life in the 2015 timeframe.

The FY-2 series is equipped with a stretched visible and infrared spin scan radiometer (S-VISSR) that scans the full Earth disk every 30 min at a spatial resolution of around 5.0 km (IR) and 1.25 km (VIS). The satellites are also equipped with SEMs to monitor the space environment where the satellites operate.

China's second-generation polar-orbiting meteorological satellite series, FY-3, began with the launch of the experimental FY-3A and FY-3B satellites on May 27, 2008 and November 5, 2010, respectively. The operational FY-3C satellite, with 11 instruments and enhanced remote sensing capabilities, was launched on September 23, 2013. The FY-3 program has a life expectancy of 15 years.

FY-4 is China's second-generation geostationary meteorological satellite series. These new generation satellites are designed with enhanced imagery scanning capabilities, desirable for monitoring small- and medium-scale weather systems. FY-4 satellites are also equipped with vertical atmospheric sounding and microwave detection capabilities to address 3D remote sensing at high altitudes. They are also able to make solar observations of extreme ultraviolet and X-ray emissions, in a bid to enhance China's space weather watch and warning capability. FY-4 satellites will come in two varieties, carrying either optical or microwave instruments. An optical satellite will carry a ten-channel 2D scanning imager, an interferometric vertical detector, a lightning imager, CCD camera, and an Earth radiation budget instrument. The satellites will produce full-disk earth images every 15 min.

8.2.2 Oceanic Satellites

The Chinese oceanic satellite program, HaiYang (HY), has been designed to detect ocean color and to support ocean bio-resource utilization, ocean pollutant monitoring and prevention, offshore resource exploration, and marine science. Eight oceanic satellites will be launched before 2020, including four ocean color satellites, two dynamic ocean environment satellites, and two ocean radar satellites.

HY-1 series satellites carry an ocean color scanner and coastal zone imager to measure global ocean color, temperature, and coastal zone dynamic change. HY-1A was an experimental mission launched on May 15, 2002 and equipped with a ten-channel ocean color scanner (3-day revisit frequency, 1.1 km nadir spatial resolution) and a four-band CCD (7-day revisit frequency, 250 m nadir spatial resolution). The follow-up satellite, HY-1B, was launched on April 11, 2007.

HY-2 series satellites measure marine environment dynamics and are used for all-weather assessments of offshore and global sea surface wind fields, sea surface height, significant wave heights, and sea surface temperature. Remote sensing payloads include a microwave scatterometer, radar altimeter, scanning microwave radiometer, and calibration microwave radiometer. The first HY-2A satellite was launched on August 16, 2011.

HY-3 will be an ocean radar satellite series used to detect ocean waves, ocean ice, and oil leaks, among other features.

8.2.3 Earth Resources Satellites

CBERS, also known as ZiYuan-1 (ZY-1), is a joint program between China and Brazil that was initiated in the 1980s. The CBERS program was designed to meet the needs of both countries and to allow them to enter the emerging market of satellite imagery. Despite an initial commitment to just two satellites, a series of missions have now been jointly built by China and Brazil. Following CBERS-1 and -2, both governments decided to expand the cooperation to include CBERS-2B, -3 and -4.

ZY-1-02C was the third satellite in the CBERS-2 series, but it was developed exclusively as a Chinese project with no Brazilian participation. It was the first high-resolution Chinese imaging mission. ZY-1-02C launched on December 22, 2011. CBERS-3 was launched on December 9, 2013 but unfortunately failed to reach its orbit. CBERS-4 was successfully launched on December 7, 2014. It will be followed by further joint China–Brazil missions.

8.2.4 China High-Resolution Earth Observation System

The China High-resolution Earth Observation System (CHEOS), also known as GaoFen (GF), is one of 16 tasks that are listed in the National Medium- and Long-term Plan for Science and Technology Development (2006–2020). The program was proposed in 2006 and approved and initiated in 2010. CHEOS is a series of small satellites for high-resolution land observation being developed between 2010 and 2020. The entire GF series of satellites will be in orbit by 2020, mostly sun synchronous around 10:30 or 06:00 except for GF-4, which will be placed into geostationary orbit. The design lifetimes are 5 years for GF-1/2 and 8 years for GF-3/4/5. China plans to launch six GF satellites between 2013 and 2016. The main goal of the GF series is to provide near-real-time observations for disaster prevention and relief, climate change monitoring, geographical mapping, environment and resource surveying, and precision agriculture.

8.2.5   Environmental Protection and Disaster Monitoring Constellation

The Environmental Protection and Disaster Monitoring Constellation mini-satellite constellation, also known as HuanJing-1 (HJ-1), is a national program to construct a network of Earth observation satellites. The overall objective is to establish an operational Earth-observing system for disaster monitoring and mitigation using remote sensing technology in order to improve the efficiency of disaster mitigation and relief.

The HJ-1 constellation is comprised of three mini-satellites, referred to as the '2 + 1 constellation'. HJ-1A and HJ-1B were launched together on September 6, 2008. HJ-1A was equipped with two optical wide view CCD cameras (WVC) and one hyperspectral imager (HSI). Both WVCs take images with four bands at 30 m spatial resolution with a swath width of 700 km. HSI measures 110–128 bands at 100 m spatial resolution with a swath of 50 km. Likewise, HJ-1B was equipped with two optical WVCs as well as one infrared multispectral scanner (IRMSS). HJ-1C was launched on November 19, 2012, equipped with an S-band synthetic aperture radar (SAR) payload.

8.2.6   The Private Sector

Twenty-First Century Aerospace Technology is a private company that has received scientific funding support primarily from the Ministry of Science and Technology of China. The company operates Beijing-1 and Beijing-2, a constellation of three small Disaster Monitoring Constellation (DMC) satellites. Beijing-1 was equipped with a 4-m panchromatic camera and a 32-m multispectral camera with a swath of 600 km. The camera, developed by Surrey Satellite Technology Ltd. (SSTL, UK), had a design lifetime of 5 years. Beijing-1 was launched on October 27, 2005. This was followed up with the launch of three Beijing-2 DMC satellites on July 11, 2015, each carrying a 1-m panchromatic camera and a 4-m multispectral camera.

The Jilin-1 series is privately owned and operated by Chang Guang Satellite Technology Co., Ltd. The series includes one optical satellite, two video satellites, and one experimental satellite. The first satellite in the series, Jilin-1A, was launched in October 2015. By 2030, it will be joined by a further 137 satellites, forming a large constellation. Jilin-1A is capable of 0.72-m panchromatic/2.88-m multispectral imaging as well as 1.12-m spatial resolution video capture.

Beijing-1/2 and Jilin-1 are market-oriented satellite programs that aim to promote the commercial service of remote sensing technology in China.

Author Biography

Jinlong Fan is a research scientist with the National Satellite Meteorological Center of the China Meteorological Administration (NSMC-CMA). He is involved in the Earth observation program of China and has contributed to the processing of global medium-resolution satellite data and its application in the fields of disaster reduction, agriculture, and the environment. He served as a seconded expert at the Group on Earth Observations (GEO) Secretariat in Geneva from 2008 to 2011 and promoted the creation of the GEO Global Agricultural Monitoring (GEOGLAM) initiative. Recently, he has been involved in numerous international activities, including serving as a Chinese partner principal investigator (PI) for the Stimulating Innovation for Global Monitoring of Agriculture (SIGMA) project, which is funded through the European Commission's FP-7 Program; a Chinese PI for the 3rd and 4th Dragon Program, a cooperation between the European Space Agency (ESA) and the Ministry of Science and Technology (MOST) of China; a member of the Committee on Earth Observation Satellites (CEOS)/Coordination Group for Meteorological Satellites (CGMS) Working Group on Climate; and a member of the expert panel on the Asia-Pacific drought mechanism, initiated by the UN Economic and Social Commission for Asia and the Pacific (ESCAP).

Chapter 9
Greenhouse Gas Observation from Space

Tatsuya Yokota and Masami Onoda

9.1 GOSAT Mission Overview

The Paris Agreement was adopted at UNFCCC COP21 in Paris, France in December 2015 as a new international framework for greenhouse gas reductions in the post-2020 period. It is a fair agreement applicable to all Parties. The Paris Agreement has a long-term objective of holding the increase in global temperatures to well below 2 °C above pre-industrial levels, and each Party shall communicate or update its Nationally Determined Contribution (NDC) every 5 years. Each party is to regularly provide information and to participate in expert reviews and a multilateral consideration of progress. A 5-yearly 'global stocktake', a review of the impact of countries' actions for the implementation of the agreement, will also take place.

CO_2 and CH_4 together account for more than 90% of the total warming effect (radiation forcing) caused by greenhouse gases (GHG) (IPCC 2013). More GHG in the atmosphere is thought to cause not only higher global average temperatures but also climatic change such as severe droughts and frequent floods, which may result in enormous damage. The Paris Agreement is a major step forward in addressing the climate change challenge. To do so, it is essential to obtain accurate information on GHG emissions on a climate zonal basis (and preferably a national basis) and to evaluate reduction measures based on this knowledge.

T. Yokota
Center for Global Environmental Research, National Institute
for Environmental Studies (NIES), 12-2 Onogawa, Tsukuba,
Ibaraki 305-8506, Japan
e-mail: yoko@nies.go.jp

M. Onoda (✉)
International Relations and Research Department, Japan Aerospace Exploration Agency,
Ochanomizu Sola City, 4-6 Kanda Surugadai, Chiyoda-Ku, Tokyo 101-8008, Japan
e-mail: onoda.masami@jaxa.jp

© The Author(s) 2017
M. Onoda and O.R. Young (eds.), *Satellite Earth Observations and Their
Impact on Society and Policy*, DOI 10.1007/978-981-10-3713-9_9

Japan's focus on creating a uniform measure from space was driven by the Kyoto Protocol adopted in December 1997 at COP3, almost 20 years before the Paris Agreement. The Greenhouse gases Observing SATellite (GOSAT) became the world's first satellite designed exclusively to observe GHG. It has been operational since launch in 2009. The mission is jointly promoted by the Japan Aerospace Exploration Agency (JAXA), the Japanese Ministry of the Environment (MOE), and the Japanese National Institute for Environmental Studies (NIES). The administrative, technical, and scientific bodies working together on this mission have been a unique and highly effective scheme for achieving its goals.

The primary objectives of GOSAT are to estimate emission and absorption of GHGs on a subcontinental scale and to assist environmental administration in evaluating the carbon balance of the land ecosystem and in making assessments of regional emission and absorption.

GOSAT measures the concentrations of CO_2 and CH_4, the two major GHGs. The technical mission targets are to (1) observe columnar CO_2 and CH_4 concentrations at 100–1000 km spatial intervals, with 1% relative accuracy for CO_2 and 2% for CH_4, during the Kyoto Protocol's first commitment period (2008–2012) and (2) reduce subcontinental scale CO_2 annual flux estimation errors by half (Kasuya et al. 2009). GOSAT complements the approximately 320 existing ground and airborne CO_2 observation points with 56,000 further points around the globe, significantly enhancing the observation network capability and providing consistent global data over a long period.

9.2   Data Products and Recent Results

GOSAT carries the Thermal and Near-infrared Sensor for carbon Observation (TANSO), which is composed of two subunits: the Fourier Transform Spectrometer (FTS) and the Cloud and Aerosol Imager (CAI). The data from FTS and CAI are processed and used together to calculate column abundances of CO_2 and CH_4 and to estimate sources and sinks as well as the three-dimensional distributions of CO_2 and CH_4 concentrations using a global atmospheric tracer transport model.

GOSAT observational data are processed at the GOSAT Data Handling Facility (DHF) of NIES and the data products are distributed to general users through the GOSAT data product distribution website (GOSAT User Interface Gateway, GUIG). The GOSAT DHF collects the specific point observation requests from qualified researchers and the observation requests of NIES and transfers them to JAXA. JAXA coordinates all observation requests to prepare the satellite operation plan.

The FTS and CAI data are received and processed into Level 1B (L1B) data at JAXA Tsukuba Space Center. These data are then transferred to the GOSAT DHF. The GOSAT DHF also collects the reference data (e.g., meteorological information) necessary for higher level processing. Using the reference data, the FTS observations are processed into column abundances (Level 2, L2), spatially interpolated monthly global distributions of column abundance (Level 3, L3), sources and sinks (Level 4A, L4A), and three-dimensional distributions of CO_2 and CH_4 (Level 4B, L4B). Reference data used for validating the products are also stored in the DHF.

GOSAT products are distributed through the GUIG. L1B data contain radiance spectra converted from raw data acquired by the satellite. The higher level products from L2–L4 store retrieved physical quantities such as the atmospheric columnar concentrations of CO_2 and CH_4. Users will be able to search and order these products using the GUIG (https://data.gosat.nies.go.jp/) by the end of 2016 or using the GOSAT Data Archive Service (GDAS, http://data2.gosat.nies.go.jp/) after January 2017.

To improve data quality, we updated the algorithm used for the estimation of XCO_2 and XCH_4 [column-averaged dry-air mole fractions (the ratio of the total amount of targeted gas molecules to the total amount of dry air molecules contained in a vertical column from the ground surface to the top of the atmosphere) for CO_2 and CH_4] and validated the retrieved values by comparing them to high-precision ground-based measurements. Using these L2 values, higher level data products such as monthly estimates of CO_2 and CH_4 regional fluxes were obtained. Based on these flux estimates, concentrations of CO_2 and CH_4 in three-dimensional space were simulated. These data have been made available to the public as GOSAT L4A (flux estimates) and L4B (three-dimensional concentration distributions). GOSAT data collected and archived for more than 6 years, can be used to map the seasonal variations and annual trends of XCO_2 and XCH_4 on regional and global scales.

The top images in Fig. 9.1 show the monthly mean GOSAT XCO_2 data gridded to a 5-degree by 5-degree mesh. The circles show GLOBALVIEW data (ground

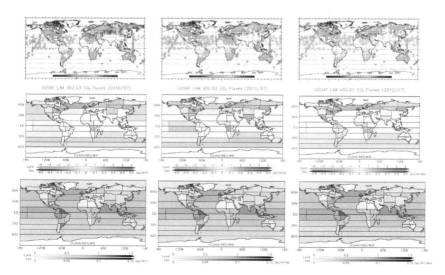

Fig. 9.1 GOSAT monthly XCO_2 mean (*top*), CO_2 flux estimates (*middle*), and CO_2 flux uncertainties (*bottom*) for July 2010, 2011, and 2012

GOSAT L4B Data Product
Model-simulated concentration
(6hr-step, 0.925 sigma-level, 2.5° × 2.5° grid)

| MM DD HH |
| 06 01 00 |

2009.6~2010.5   2010.6~2011.5   2011.6~2011.10

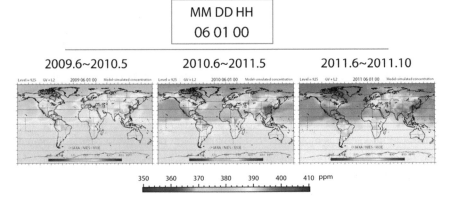

350   360   370   380   390   400   410 ppm

Fig. 9.2 Examples of GOSAT L4B data—model-simulated CO_2 concentrations for the same hour, date, and month in 2009, 2010, and 2011. Color denotes CO_2 concentration. 0.925 sigma-level represents about 800 m in the altitude of the mid-latitude atmosphere

observation, 212 sites). With this input, the middle images are generated (monthly flux estimates) and the bottom images show flux uncertainties (GOSAT L4A).

Figure 9.2 shows the L4B data product, which is the result of an atmospheric tracer transport model simulation based on the flux distribution (L4A) estimated from the ground-based and GOSAT-based concentration data. L4B products store global concentrations using a 2.5-degree mesh in intervals of 6 h at 17 vertical levels, ranging from near the surface to the top of the atmosphere.

MOE, NIES, and JAXA issued a press-release on December 4, 2014 stating that GOSAT archive data has the potential to detect the origin of increased CO_2 concentrations. These analyses have progressed and have been performed for the Tokyo metropolitan area and other major cities around the world. The results, announced on September 1, 2016, demonstrated for the first time the possibility of using satellite observations to monitor and verify the emissions reported by countries, even at relatively small scales.

Figure 9.3a shows areas where high concentrations of anthropogenic CO_2 emissions were observed (average from June 2009 to December 2014). The color represents concentration. Figure 9.3b shows the correlation between the satellite data and inventory estimates for Japan.

These results demonstrate that satellite measurements have the potential to be used for Measurement, Reporting, and Verification (MRV)—especially verification for multilateral agreements—in combination with ground-based, airborne, and other measurements. For such purposes, it is critical that data are free and open.

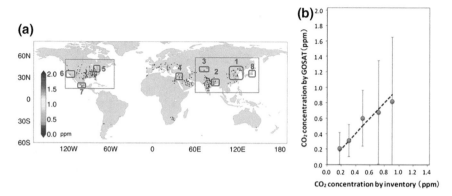

Fig. 9.3 **a** Distribution of anthropogenic CO_2 concentrations higher than 0.1 ppm between June 2009 and December 2014, as estimated from observational data acquired by GOSAT. Red squares represent typical high emission areas (megacities). **b** Relationship between GOSAT estimates of anthropogenic concentrations and inventories in Japan

9.3   The Way Forward: GOSAT-2

GOSAT-2 is scheduled for launch in 2018. Developed jointly by JAXA, MOE, and NIES, GOSAT-2 is a continuation of the GOSAT mission with upgraded observation capabilities to meet the increased information demands of, for example, the Paris Agreement. GOSAT-2/TANSO-FTS-2 will also have full pointing capability, allowing cloud avoidance and targeted observations of large emission sources. GOSAT-2 will be able to observe carbon monoxide as a new observation target and aerosol pollutants such as PM2.5 or black carbon in the atmosphere.

Japan has submitted to the IPCC a proposal for guidelines on the use of satellite data for verifying or validating carbon inventories, which may contribute to worldwide efforts to monitor the state of the global carbon cycle and the effect on the Earth's atmosphere and to help countries achieve their obligations. While satellite data cannot be expected to immediately replace existing methods, the possibility of applying satellite data to national inventories, or in the report and review process, is becoming more realistic.

As the sink and source distribution of CO_2 can be evaluated by the inhomogeneity of atmospheric concentration using diagnostic models, for accurate evaluation it is essential to enhance the current observation network, especially in areas where observations are sparse. Filling the existing spatial observation gaps would improve the quality of information and understanding of the long-term and general status of climate change, thus reducing the uncertainty of the scientific basis of treaties.

Over the long term, satellite observations might contribute to evaluating the sink and source distribution of CO_2 at a precision satisfactory to verify treaty effectiveness over long timeframes.

Satellite missions should aim to provide observations over a long period, so continuity is essential. Total column CO_2 observation missions are planned in a harmonized manner, and include GOSAT (Japan), OCO-2 (NASA), TanSat (China), GOSAT-2 (Japan), Microcarb (CNES), and possibly Carbonsat (ESA) and GOSAT-3 (Japan). A combination of measurement platforms (including in situ) is important to understand the status and change of GHG distribution and to estimate carbon sources and sinks.

Further work is needed to bridge the gap between observation methods and the policy framework. We need not only to enhance measurement accuracies, but also to develop and improve models that identify anthropogenic emissions. The results may be applied to the verification of national inventories or to estimating the effectiveness of measures taken, thus contributing to future treaty amendments and new institutional negotiations. By addressing the gaps between observations and the institutional frameworks, worldwide satellite missions may serve a significant role in the Paris Agreement and beyond.

Reference

Kasuya M et al (2009) Greenhouse Gases Observing Satellite (GOSAT) Program Overview and Its Development Status. https://www.jstage.jst.go.jp/article/tstj/7/ists26/7_ists26_To_4_5/_pdf. Accessed 7 Dec 2016

Author Biographies

Tatsuya Yokota received his Ph.D. in measurement and information systems engineering from The University of Tokyo in 1987. Since 1981, he has been working for the National Institute for Environmental Studies (NIES), Japan. His specialty is information processing and atmospheric remote sensing. He was engaged in several atmospheric satellite remote sensing projects in Japan, including ILAS, ILAS-II, and SOFIS, for polar ozone layer monitoring. He served as the project leader of the Greenhouse Gases Observing Satellite (GOSAT, *Ibuki*) at NIES and as the head of the Satellite Remote Sensing Research Section at the Center for Global Environmental Research (CGER) (2006–2016). Since April 2016, he has been a NIES fellow and the acting leader of the NIES GOSAT project.

Masami Onoda is currently the U.S. and multilateral relations interface at the International Relations and Research Department of the Japan Aerospace Exploration Agency (JAXA). As an academic, she is fellow of the Institute of Global Environmental Strategies (IGES) and she is also engaged in the private sector as an advisor to the Singapore-based space debris start-up Astroscale Pte. Ltd. since its foundation in 2013. From 2009 to 2012, Dr. Onoda was a scientific and technical officer at the intergovernmental Group on Earth Observations (GEO) Secretariat in Geneva, Switzerland. From 2003 to 2008, while pursuing her graduate studies, she was invited to the JAXA Kansai Satellite Office in Higashiosaka as a space technology coordinator to support technology transfer to SMEs for the small satellite project SOHLA-1. From 1999 to 2003, she worked in the field of Earth observations at JAXA (then NASDA), serving on the Secretariat of the Committee on Earth Observation Satellites (CEOS). In 1999, she was seconded to the UN Office for Outer Space Affairs (UNOOSA) for the organization of the UNISPACE III conference. She holds a Ph.D. in global environmental studies (2009) and a master's degree in environmental management (2005), both from the Kyoto University Graduate School of Global Environmental Studies. Her undergraduate degree is in international relations from The University of Tokyo.

Chapter 10
Japanese Satellite Earth Observation: Status and Policy Issues

Teruyuki Nakajima

10.1 Introduction

The global Earth observation satellite system has continually progressed and evolved, with the development of international meteorological frameworks such as the World Weather Watch (WWW) and Global Atmospheric Research Programme (GARP) in the 1960s; with increased awareness of the global climate and environmental problems, thanks to NASA's Mission to Planet Earth, the European Space Agency's (ESA) ENVISAT, and JAXA's ADEOS/ADEOS-II satellite programs; the Global Climate Observing System (GCOS) in the 1990s; and the implementation of the UN Framework Convention on Climate Change (UNFCCC), the intergovernmental Group on Earth Observations (GEO), and others from the 2000s. This progress is linked to various policy mechanisms, both in terms of program development and the utility of remote sensing results and applications. This article discusses some important aspects of current issues related to Earth observation policies.

10.2 Development of National Satellite Earth Observation Programs

JAXA has operated 20 Earth observation satellites in total, four of which (GOSAT, GCOM-W, GPM, and ALOS-2) remain operational in 2016. After enactment of the Basic Plan on Space Policy, Japanese national space utilization planning has been governed by the Space Policy Committee (SPC) of the Cabinet Office. The SPC has

T. Nakajima (✉)
Earth Observation Research Center (EORC), Japan Aerospace Exploration Agency (JAXA),
2-1-1 Sengen, Tsukuba, Ibaraki 305-8505, Japan
e-mail: nakajima.teruyuki@jaxa.jp

© The Author(s) 2017
M. Onoda and O.R. Young (eds.), *Satellite Earth Observations and Their Impact on Society and Policy*, DOI 10.1007/978-981-10-3713-9_10

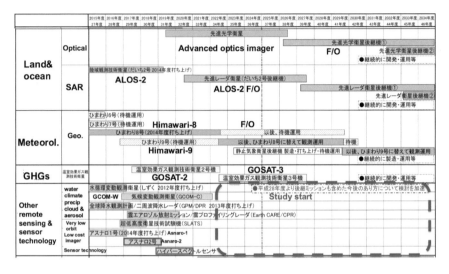

Fig. 10.1 National Earth observation work plan

identified seven new governmental priorities: positioning, remote sensing, communication/broadcasting, space science, human space activity, space solar power, and space transportation. The SPC also determined the national Earth observation work plan (CAO 2016) shown in Fig. 10.1. Geostationary meteorological satellites (the Himawari series), greenhouse gas observation satellites (the GOSAT series), and advanced optical and SAR imaging satellites for Earth surface monitoring have been approved. There is, however, currently a serious gap in capability after 2022 for other remote sensing categories, though it should be noted that the SCP will begin investigating a plan for a GCOM-W follow-on mission from the 2016 fiscal year.

In contrast, other countries have been very active in planning new Earth observation missions. There are now more than 150 satellites planned to contribute to global climate change and environmental initiatives, such as the European Commission (EC) Copernicus program, the UN Global Framework for Climate Services (GFCS), and others.

In this situation, it is important to have a strategic vision and satellite build program for the 2020–2040 period. A new trend for this period will be a close collaboration between space agencies and weather/climate agencies, illustrated by the collaboration between ESA and the European Organisation for the Exploitation of Meteorological Satellites (EUMETSAT) in the Copernicus program, and the collaboration between NASA and the National Oceanic and Atmospheric Administration (NOAA) in the Joint Polar Satellite System (JPSS) program. China and Korea are also very actively planning capabilities for this period.

It is therefore a matter of urgency for Japan to reinforce its national Earth observation satellite planning to support the growing challenge of understanding and contributing to solutions for global climate change and other related environmental problems. There has been no high-level national platform in the Cabinet Office for professionals to discuss and manage such huge requirements and to chart the path for the development of capabilities. A subcommittee for Earth observation under the SPC should be established.

Another important activity for enhancing national Earth observation planning in Japan is to strengthen the bottom-up planning by academic organizations. From 2014, the Science Council of Japan (SCJ) established the Earth Observation Sub-Committee under the Earth and Planetary Science Committee. In 2013, the Remote Sensing Society of Japan started a Task Force Sub-Committee for discussion of future space development. The Task Force now comprises more than 20 societies/associations. The SCJ and Task Force discussions and commendations indicate that national Earth observation satellite missions have contributed to research and social services and that Japan should continue to develop various applications for cutting-edge science and service of society. It is important to also note the high citation index and H-index of scientific papers related to Earth observation satellites.

Figure 10.2 shows the 'dream' roadmap toward 2050 proposed by the SCJ (2014). It is significant to find that the research community identifies a need for Earth observations across all periods to inform scientific investigations and provide services for society, in areas such as cloud and climate systems, as well as disaster prevention, among others. Related issues for making observations useful are development of various systems to enable maximum use of observation data and models for climate projections, accurate forecasting of severe weather phenomena for disaster prevention, and other applications, as raised also by the International Council for Science (ICSU) Future Earth Initiative. One of the future directions is to build combined applications of Earth observation and Earth system modeling, including human dimensions and associated High Performance Computer Infrastructure (HPCI).

There are many requirements for sustained Earth system observations from space, necessitating strong international coordination through groups such as the Committee on Earth Observation Satellites (CEOS), GEO, and the World Meteorological Organization (WMO). At the same time, space agencies must strive to reduce costs, extend mission lifetimes, and seek effective collaboration with the business sector. Such efforts can help humanity achieve affordability of sustained and critical observations of the Earth from space, for both research and societal applications.

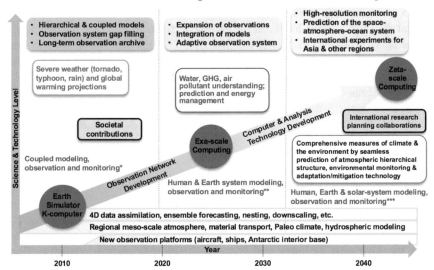

Fig. 10.2 The 'dream' 2050 roadmap of the SCJ for atmosphere and hydrosphere sciences (SCJ 2014). *Coupled modeling includes problems related to mesoscale weather and cloud system resolution; aerosols, chemistry, turbulence/cumulous/gravity wave parameterizations; whole atmospheric modeling; ocean hierarchical structure modeling; and water cycle coupling. Observation and monitoring for this period includes Earth surface observation networks [weather stations, mesosphere-stratosphere-troposphere (MST) radar, marine radar, buoys]; aircraft and ship-based monitoring (including radioactive material); satellite-based measurements (of clouds, winds, temperature, water vapor, precipitation, greenhouse gases); long-term reanalysis of climate data; intensive regional observations (cloud radar, sondes, etc.); Asian biodiversity observations; Antarctic and Arctic ice core analysis, next-generation core technology; and measurements of solar conditions. **Human and Earth system modeling includes cloud system resolution/modeling; cloud microphysics, radiation, boundary layer turbulence processes, wave hierarchical structure, material transport and diffusion; ocean inter-process interaction; ecosystem–water cycle interaction; and information provision. Observation and monitoring required for this period includes multi-element and comprehensive observation networks (aircraft, radar, multi-function LIDAR, sondes, ground stations); adaptive observations (severe weather, transboundary pollution); global and coastal ocean observations (tsunami, radar profiling, buoys, environmental sample analysis); improvement of global low-Earth orbit/geostationary satellite observations (of clouds and aerosols, greenhouse gases, the water cycle, marine life, vegetation); global ice core sampling; middle atmosphere monitoring (of the ozone layer, noctilucent clouds); all water cycle processes; a geospace observation system; and a solar-terrestrial correlation measurement system. ***Human, Earth and solar-system modeling includes cloud and turbulence resolution/modeling and hydrology, marine flux, and solar activity process modeling. Observations required in this period include: those related to understanding geospace–atmosphere–human activity interactions; monitoring using networks of meteorological, hydrological, and ecological systems with adaptive pluralistic monitoring and operation; steady operation of satellite observations of the water cycle, climate change, and whole atmospheric layers; monitoring of abrupt and abnormal marine phenomena; comprehensive management of spatial and temporal variabilities of the marine ecosystem and resources; and Antarctic grid drilling and planetary ice sheet drilling

10.3 Development of Earth Observation Products and Applications

Mitigation of and adaptation to the changing climate and environmental problems will continue to be a priority application of national Earth observation programs. The EC's Copernicus program, which aims to secure operational Earth observation satellite data for climate and other services, is a good example. Another feature is the emergence of open and free Earth observation data policies that promote climate services and businesses for social applications.

Japanese efforts to actively use Earth observation data to address societal problems are not strong. In response, JAXA has enhanced the multi-satellite research activities of the Earth Observation Research Center (EORC) in important areas such as disaster prevention, ocean monitoring, water cycle/resource management, atmospheric monitoring, infrastructure monitoring, climate system/radiative process studies, ecosystems, agriculture, and public health (Fig. 10.3). Programs use data from the Advanced Himawari Imager (AHI) on board the world's first third-generation geostationary satellite, Himawari-8, thanks to a collaboration with the Japan Meteorological Agency (JMA).

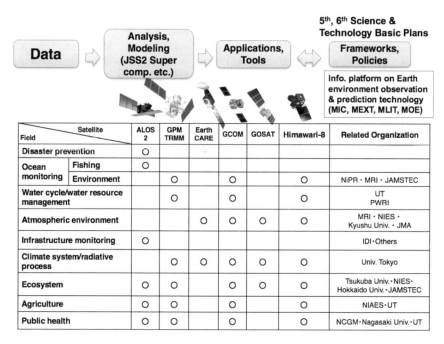

Field \ Satellite	ALOS 2	GPM TRMM	Earth CARE	GCOM	GOSAT	Himawari-8	Related Organization
Disaster prevention	O		-				
Ocean monitoring — Fishing	O						
Ocean monitoring — Environment		O		O		O	NiPR · MRI · JAMSTEC
Water cycle/water resource management		O		O		O	UT PWRI
Atmospheric environment			O	O	O	O	MRI · NIES · Kyushu Univ. · JMA
Infrastructure monitoring	O						IDI·Others
Climate system/radiative process		O	O	O	O	O	Univ. Tokyo
Ecosystem	O	O		O	O	O	Tsukuba Univ.·NIES· Hokkaido Univ.·JAMSTEC
Agriculture	O	O		O		O	NIAES·UT
Public health	O	O		O		O	NCGM·Nagasaki Univ.·UT

Fig. 10.3 The crosscutting research fields of JAXA's EORC

Monitoring sea surface temperature (SST) and sea ice coverage using AMSR-2 (GCOM-W imaging radiometer) is useful for weather forecasting and climate projections. The Global Satellite Mapping of Precipitation (GSMaP) initiative produces a high-precision, high-resolution global precipitation map using data from multiple satellites, including microwave imagers and the core radar instrument of the Global Precipitation Measurement (GPM) mission. Satellite monitoring systems have proven especially useful during serious natural disasters, such as the flooding of the Kinu river, increased volcanic activity in the Hakone mountains, and the Kumamoto earthquake. As a result, Japanese society is increasingly interested in satellite monitoring. It is important to recognize that improved communication and data dissemination networks have amplified this interest.

Long-term time series of emissions of various short-lived atmospheric components can be obtained by integrating atmospheric chemistry transport models with such space technologies as Ozone Monitoring Instruments (OMI), Microwave Limb Sounders (MLS), Tropospheric Emission Spectrometers (TES), and Measurements of Pollution In The Troposphere (MOPITT) (Miyazaki et al. 2015). Column loadings of Long-Lived Greenhouse Gases (LLGHGs) have also been observed by GOSAT and OCO-2. This has increased our ability to accurately monitor global LLGHG emissions from human activities and to take collective action for their reduction through UNFCCC mechanisms such as Reducing Emissions from Deforestation and Forest Degradation in Developing Countries (REDD+) for carbon monitoring and attribution. A recent study (Turner et al. 2015), for example, showed that the U.S. Environmental Protection Agency (EPA) inventory of methane was underestimated in the North American continent. Shortening the cycle of data assimilation and inversion will change the strategy for controlling the enforcement of LLGHG emissions and air pollutants subject to control by the UNFCCC and the Climate and Clean Air Coalition (CCAC). Such accelerated construction of the emission inventory can yield new strategies for building co-benefits for climate and public health by combining monitoring data and atmospheric chemistry modeling.

Observations using the L-band SAR instruments PALSAR and PALSAR-2 have enabled JAXA to undertake accurate global-scale forest monitoring. One example, 1200 km^2 of observations over Brazil from 2009 and 2014 (by ALOS and ALOS-2, respectively) revealed a total deforestation area of 25 km^2 for the region of interest. ALOS-2 has also observed the crustal movement of the Hakone area caused by volcanic activity and landslides from the earthquakes in Nepal and Kumamoto. Data have been provided to local governments and authorities (in support of disaster response actions) through national channels and the International Charter on Space and Major Disasters.

Recent progress in SAR technology has even made it possible to monitor subsidence of reclaimed land. Time series interferometry quantitatively visualizes deformation that reveals subsidence area over monitored structures. The results show good agreement between interferometry and ground measurement data within 11.5 mm root mean square error. Such a capability can lead to new infrastructure management and city planning services. This feature is especially important for Japan, as authorities work to investigate and replace old infrastructure developed in the high growth period of the 1960s and 1970s.

10.4   Conclusions

There is a global trend of expanding Earth observation satellite program planning for the 2020–2040 period, but Japanese plans for this period are not well established. The space work plan in the category of other remote sensing is not yet determined beyond 2022 as shown in Fig. 10.1. This category includes important satellites from which useful data are obtained for cloud microphysics, Earth radiation budget, vertical stratification of atmospheric constituents, and others to improve Earth system and environmental models. It is therefore necessary to enhance Japan's national Earth observation program planning process to prepare and contribute to solutions for changing global climate and environmental problems. At the same time, Japan should further connect satellite Earth observation with advanced modeling and societal services in order to attain maximum use of Earth observation systems. Through this connection, comprehensive satellite data analyses will become more feasible and useful, not only for short-term weather forecasting and long-term global warming projection, but also for medium-term forecasting such as for seasonal weather change and agriculture.

References

CAO (2016) Uchuu kihon keikaku (Basic Space Plan). http://www8.cao.go.jp/space/plan/keikaku.html. Accessed 7 Dec 2016
Miyazaki K et al (2015) A tropospheric chemistry reanalysis for the years 2005–2012 based on an assimilation of OMI, MLS, TES, and MOPITT satellite data. Atmos Chem Phys 15:8315–8348
SCJ (2014) Dream roadmap 2014. http://www.scj.go.jp/ja/info/kohyo/pdf/kohyo-22-h201-3-4.pdf. Accessed 7 Dec 2016
Turner AJ et al (2015) Estimating global and North American methane emissions with high spatial resolution using GOSAT satellite data. Atmos Chem Phys 15:7049–7069

Author Biography

Teruyuki Nakajima has greatly contributed to studies of the earth radiation budget, climate impacts of clouds and aerosols, and air pollution, including from the Fukushima Nuclear Power Station accident, by building observation systems and models for atmospheric constituents. He has served and promoted Earth science as the secretary general of the International Association of Meteorology and Atmospheric Sciences (IAMAS) and has held positions as a member of the Science Council of Japan, president of the Atmospheric and Hydrospheric Sciences Section of the Japan Geoscience Union, and a member/officer of the World Climate Research Programme (WCRP)/Joint Science Committee. He also contributed to the Intergovernmental Panel on Climate Change (IPCC) assessments of global warming as a lead author for the Third and Fifth IPCC Assessment Reports.

Part IV
International Initiatives and Studies

Chapter 11
The New 10-Year GEOSS Strategy for 2016 and Beyond

Barbara Ryan and Osamu Ochiai

11.1 Introduction to GEO and GEOSS

Global environmental change, and its impact on all aspects of society, is one of the primary challenges facing humanity, even more so today than in 2003 when governments and international organizations committed to a vision of a future wherein decisions and actions for the benefit of humankind are informed by coordinated, comprehensive, and sustained Earth observations. In 2005, the first concrete step toward achieving that vision was taken with the establishment of the Group on Earth Observations (GEO), whose primary goal is to create the Global Earth Observation System of Systems (GEOSS) (Fig. 11.1, 11.2).

GEO is a partnership comprising governments and international bodies with a mandate in, and/or use of Earth observations. GEO now consists of 104 Members, 106 Participating Organizations, and 12 Observers.

11.2 GEO Strategic Plan 2016–2025: Implementing GEOSS

Recognizing the need for further collective effort to foster the use of Earth observation resources to their fullest extent, GEO Ministers extended the mandate of GEO for a second decade at the Geneva Summit in January 2014.

B. Ryan (✉) · O. Ochiai
Group on Earth Observations, 7 Bis, avenue de La Paix, Case Postale 2300,
1211 Geneva 2, Switzerland
e-mail: bryan@geosec.org

O. Ochiai
e-mail: oochiai@geosec.org

Fig. 11.1 GEOSS

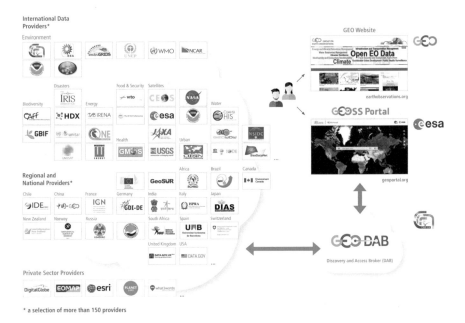

Fig. 11.2 GEOSS promotes broad open data policies and enables interoperable systems

The *GEO Strategic Plan* 2016–2025: *Implementing GEOSS* was prepared in response to this renewal and builds upon the strong foundation of GEO and its proven successes. The Plan identifies improvements in areas highlighted in the *Geneva Declaration*, including strengthening the Societal Benefit Areas (SBAs); engaging more broadly with stakeholders, the United Nations (UN), and the private sector; establishing a robust, steady resourcing mechanism within the voluntary framework of GEO; and identifying new opportunities.

11.2.1 Societal Challenges and Opportunities

When integrated, Earth observations from diverse sources—including satellite, airborne, in situ platforms, and citizen observatories—provide powerful tools for understanding the past and present conditions of Earth systems, as well as the interplay among them.

GEO is facilitating the development of solutions to societal challenges within the following SBAs by mobilizing resources—including observations, science, modeling and applications—to enable end-to-end systems and deliver services for users.

- Biodiversity and Ecosystem Sustainability
- Disaster Resilience
- Energy and Mineral Resources Management
- Food Security and Sustainable Agriculture
- Infrastructure and Transportation Management
- Public Health Surveillance
- Sustainable Urban Development
- Water Resources Management

Climate change and its impacts cut across all SBAs. GEO will supply the requisite Earth observations in support of effective policy responses for climate change adaptation, mitigation, and other impacts across the SBAs.

11.2.2 Stakeholder Engagement

The value of Earth observation data is fully realized when it is transformed into useable knowledge and information to address societal needs. GEO will therefore convene key stakeholders across the provider–user spectrum to co-design a process to systematically identify and document Earth observation needs to address specific problems within the scope of the SBAs.

GEO will build stronger relationships with complementary global Earth observation organizations, including UN Agencies that are already Participating Organizations, as well as other national, regional, and global entities, particularly in regard to the important role Earth observations play in support of measuring, monitoring, and achieving the United Nations Sustainable Development Goals (SDGs) (Fig. 11.3).

GEO also offers unique information and engagement opportunities for the private sector and international development banks to serve their needs in areas such as agriculture, transportation, resource extraction, and insurance. In turn, GEO benefits from the participation of the private sector through access to new types of data, diverse capabilities and new technologies, and broader community networks.

Fig. 11.3 Earth observations and geospatial information contribute to UN SDG indicators and targets both directly and indirectly

Further, development banks can offer a unique opportunity for GEO to engage directly with developing countries. Given these mutual benefits, the private sector and development banks have the capacity to be key contributors to making substantive progress towards achieving GEO's Strategic Objectives.

11.2.3   Core Functions

GEO will implement a set of Core Functions essential for the realization of its Strategic Objectives. These Core Functions are accompanied by specific, measurable, and achievable targets designed to enable effective implementation, including the monitoring of progress and evaluation of achievements.

The Core Functions for GEO are:

- Identifying user needs and addressing gaps in the information chain;
- Sustaining foundational observations and data;
- Fostering partnerships and mobilizing resources;
- Advancing GEOSS and best practices in data management and sharing;
- Implementing sustained global and regional services; and
- Cultivating awareness, building capacity, and promoting innovation.

11.2.4   Implementation Mechanisms

Because of its broad intergovernmental membership and variety of contributing organizations, GEO is able to assemble and coordinate expertise from across numerous disciplines and communities. GEO uses this convening power to bring together the unique combinations of partners required to address societal challenges faced by communities across the globe at every scale, from individuals to countries and continents, drawing on comprehensive, coordinated, and sustained Earth observations. This mechanism (Fig. 11.4) will not only allow the establishment of a robust, steady resource within the framework of GEO, but also identify new

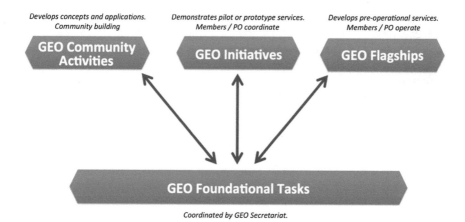

Fig. 11.4 GEOSS implementation mechanism

opportunities for GEO. The following types of activities contribute to GEO's Strategic Objectives in various ways:

GEO Community Activities allow stakeholders to cooperate flexibly in a bottom-up fashion and with a low initiation cost;

GEO Initiatives allow Members and Participating Organizations to coordinate their actions and contributions towards a common objective within an agreed, yet flexible framework;

GEO Flagships allow Members and Participating Organizations with a policy-relevant mandate to spin-up a dedicated operational service serving common needs and/or well-defined user groups; and

GEO Foundational Tasks allow GEO to implement selected, often enabling, tasks to achieve GEO Strategic Objectives. These include coordination actions, gap analyses, the implementation of technical elements for accessing GEOSS, and other routine operations of the GEO Secretariat. Thus, they provide important support functions to Flagships, Initiatives, and Community Activities.

11.2.5 Governance Structure

In order to successfully achieve its Vision and Strategic Objectives and to develop all related functions, GEO will rely on governance arrangements as set out in the GEO Rules of Procedure. Ministers of the GEO Members, meeting periodically, will provide the political mandate and overall strategic direction for GEO.

Plenary: The highest decision-making body of GEO is composed of Principals at the senior-official level, or their designated Alternates, representing GEO Members and Participating Organizations.

Executive Committee: An Executive Committee, composed of representatives of GEO Members, provides the strategic leadership for GEO activities when the Plenary is not in session.

Programme Board: A Programme Board, composed of persons nominated by GEO Members and Participating Organizations, will support the development and implementation of GEO activities. It also examines proposed Implementation Plans for GEO Initiatives and Flagships.

Secretariat: A Secretariat, led by a Director and accountable to Plenary and the Executive Committee, will facilitate and support GEO activities.

11.3   Moving Forward

As GEO enters the next decade, it will focus on addressing societal challenges facing humankind by advocating the value of Earth observations; engaging with key stakeholders, including the private sector and development banks; and delivering data, information, and knowledge that is critical to informed decision-making.

Author Biographies

Barbara J. Ryan is secretariat director of the intergovernmental Group on Earth Observations (GEO) located in Geneva, Switzerland. In this capacity, she leads the Secretariat in coordinating the activities of 103 Member States and the European Commission and the 106 Participating Organizations that are integrating Earth observations so that informed decisions can be made across eight societal benefit areas (SBAs), namely, Biodiversity and Ecosystem Sustainability; Disaster Resilience; Energy and Mineral Resources Management; Food Security and Sustainable Agriculture; Infrastructure and Transport Management; Public Health Surveillance; Sustainable Urban Development; and Water Resources Management. Before becoming GEO Director in July 2012, Ryan served as director of the World Meteorological Organization (WMO) Space Programme with responsibility for coordinating space-based observations to meet the needs of WMO members in the topical areas of weather, water, climate, and related natural disasters. Before joining WMO in October 2008, she was the associate director for geography at the U.S. Geological Survey (USGS) in Reston, Virginia, where she had responsibility for the Landsat, remote-sensing, geography and civilian mapping programs of the agency. It was under her leadership that implementation of the Landsat data policy was reformed to release all data over the Internet at no additional cost to the user—an action that has resulted in the global release of more than 40 million Landsat scenes to date and generated significant economic returns globally.

Osamu Ochiai is a scientific and technical officer at the Group on Earth Observations (GEO) Secretariat in Geneva, Switzerland. He was seconded as an expert in the area of the Global Earth Observation System of Systems (GEOSS) Architecture and Data Management from the Japanese Aerospace Exploration Agency (JAXA). He has spent more than 15 years working in the field of satellite-based Earth observation data and information systems, as well as various data utilization and application programs at JAXA. He has also been involved with the Committee on Earth Observation Satellites (CEOS) since 1998. He holds a degree in geophysics and a M.Sc. from the University of Hokkaido, Japan.

Chapter 12
The Value of Global Earth Observations

Michael Obersteiner, Juraj Balkovič, Hannes Böttcher, Jetske A. Bouma, Steffen Fritz, Sabina Fuss, Peter Havlik, Christine Heumesser, Stefan Hochrainer, Kerstin Jantke, Nikolay Khabarov, Barbara Koch, Florian Kraxner, Onno J. Kuik, Sylvain Leduc, Junguo Liu, Wolfgang Lucht, Ian McCallum, Reinhard Mechler, Elena Moltchanova, Belinda Reyers, Felicjan Rydzak, Christian Schill, Christine Schleupner, Erwin Schmid, Uwe A. Schneider, Robert J. Scholes, Linda See, Rastislav Skalský, Alexey Smirnov, Jana Szolgayova, Zuzana Tarasovičová and Hong Yang

Humankind has never been so populous, technically equipped, and economically and culturally integrated as it is today. In the twenty-first century, societies are confronted with a multitude of challenges in their efforts to manage the Earth system. These global challenges range from multi-hazard disasters and new infectious diseases to basic food and energy security on a warming planet. Dealing with such a confluence of possible global-scale failures involves highly complex planning, coordination, and international cooperation (Walker et al. 2009); this will only progress effectively and efficiently if private and public policies are based on reliable information and sound science. Particularly distressing in this context are the continued deficiencies in basic global observations and information-processing infrastructure to monitor and document many of the important ongoing global changes with sufficient accuracy. Information paucity remains an obstacle to understanding major Earth system processes. In this decision-making context, collective efforts to manage the Earth system are in danger of being erratic and the result of competing interests rather than based on decisions informed by robust scientific analysis. While international conventions on the environment and security are today the main drivers of global Earth observations, the resulting space and in situ observing instruments are far from optimally structured, deployed, or used when considered in light of the value of the decisions that are at stake.

M. Obersteiner (✉) · J. Balkovič · H. Böttcher · J.A. Bouma · S. Fritz · S. Fuss · P. Havlik ·
C. Heumesser · S. Hochrainer · K. Jantke · N. Khabarov · B. Koch · F. Kraxner · O.J. Kuik ·
S. Leduc · J. Liu · W. Lucht · I. McCallum · R. Mechler · E. Moltchanova · B. Reyers ·
F. Rydzak · C. Schill · C. Schleupner · E. Schmid · U.A. Schneider · R.J. Scholes · L. See ·
R. Skalský · A. Smirnov · J. Szolgayova · Z. Tarasovičová · H. Yang
International Institute for Applied Systems Analysis, Schlossplatz 1, Laxenburg 2361, Austria
e-mail: oberstei@iiasa.ac.at

© The Author(s) 2017   137
M. Onoda and O.R. Young (eds.), *Satellite Earth Observations and Their
Impact on Society and Policy*, DOI 10.1007/978-981-10-3713-9_12

Currently the most pertinent example of high-impact decision-making informed by science that would greatly benefit from an enhanced observing system is connected to the ongoing controversy around the interpretation of the mandate of the UN Framework Convention on Climate Change (UNFCCC) for *stabilization of greenhouse gas concentrations in the atmosphere at a level that would prevent dangerous anthropogenic interference with the climate system.* Although the Global Climate Observing System (GCOS) is already considered to be advanced in its implementation, the benefit of reducing uncertainty in climate predictions by even better informed models through an improved GCOS justifies further investment. The incremental annually recurring cost of implementing the GCOS is estimated to be in the range of 600–700 million USD, which can be compared to the average annual incremental climate mitigation cost for the next 20 years in the energy system, amounting to some 300 billion USD per annum (Rao 2009).

At the two UN Earth Summits, it was realized that complex Earth processes can be adequately measured to support environmental decision-making only by linking and coordinating current observing systems. Since then, a number of Earth observation summits have been held, resulting in the establishment of the intergovernmental GEO at the third occasion. GEO provides the platform for coordinating observation strategies and investment in support of decision-making in nine SBAs (see Sect. 4.1.2.1).

Prioritization of coordinating actions and investments to build a joint GEOSS necessitates an integrated assessment of the prospective economic, social, and environmental benefits. We have therefore developed methodologies and analytical tools—following a benefit chain concept (GEOBENE 2016)—and applied them to assess the societal benefits of investments in improving the GEOSS across the nine SBAs. The basic idea is that the costs incurred by an incremental improvement in the observing system—including data collection, interpretation, and information sharing—will result in benefits through information cost reduction or better informed decisions. The resulting incremental societal benefit is judged against the incremental cost of production. Since in many cases there are large uncertainties in the estimation of costs and particularly the benefits, it may not be possible to express them in comparable monetary terms. Therefore, order-of-magnitude approaches and a qualitative understanding of the shape of the cost–benefit relationships can help guide investment decisions.

There are generally two source categories for cost reductions in information delivery from building the GEOSS. The first relates to cost reduction from economies of scale of a global or large observing system vis-à-vis the currently prevailing patchwork system of national or regional observing systems. For example, the costs of national forest carbon assessments aimed at policies of avoided deforestation typically amount to 100–500 USD per km^2, yielding carbon stock estimation uncertainties of between 10–20%. A similar precision achieved by one consistent global forest observatory could be realized at much lower costs of some 10–100 USD per km^2 (Böttcher et al. 2009).

The second source of cost reduction from GEOSS relates to economies of scope, which emerge when observing systems are combined. The Geo-Wiki Project

(Fritz et al. 2009) combines human sensors (a global network of volunteers) with satellite images to improve global land information. Substantially improved assessment of land resources are of particular importance, for example, for the currently hotly debated estimation of indirect land use effects of biofuel policies (Searchinger et al. 2009). We estimated that the opportunity cost of avoiding deforestation, thereby minimizing indirect land use effects, to differ on the order of 50% depending on which estimates of cropland availability are used, as taken from different state-of-the-art yet insufficiently accurate land cover maps (Fritz et al. 2009).

Economies of scale and scope are not only about leveraging cost reduction, but also accrue from increased benefit generation from integrating observing systems. Quantifying benefits, often of a "public good" nature, proves a significant challenge. In a local case study in the Little Karoo of South Africa, we investigated the benefits of improved land cover information, of the type to be expected from GEOSS, for local-scale ecosystem service monitoring for the parameters presented in Fig. 12.1. Using precolonial ecosystem service levels as a reference point, the assessment demonstrated substantial differences in current ecosystem service levels when measured using local-scale ground-truthed data associated with GEOSS systems versus national-scale non-ground-truthed data. The former finds substantial declines in ecosystem service levels of between 18–44% of precolonial levels, concurring with other local studies that highlight the extreme decline in ecosystem services in the Little Karoo region, while the latter paints a much rosier—but

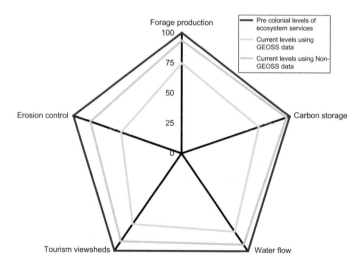

Fig. 12.1 Measuring changes in ecosystem services from precolonial times until the present day using GEOSS (local scale, ground-truthed) and non-GEOSS (national scale, non-ground-truthed) scenarios of data availability. The current levels are reflected as a percentage of the precolonial levels for each ecosystem service. Higher levels of ecosystem service degradation are identified in the GEOSS scenario revealing more degradation from precolonial levels (100%). Improved information on the actual degree of degradation (GEOSS levels) improves decisions and generates benefits such as avoided losses from floods. Data are extracted from Reyers et al. (2009)

incorrect—picture of almost intact systems providing ecosystem services at near precolonial levels (10–15% declines). Applying the benefit chain concept, we compare the costs of GEOSS-type data of some 12,000 USD to the incremental benefits of more accurate information on current ecosystem service condition. However, due to the limited development of procedures for quantifying the economic benefits of improved environmental information, the case study currently relies on proxies such as the costs of wrong decisions (including restoration costs of approximately 1100 USD per hectare) and flood costs (which in 2006 totaled 40 million USD damage in this region).

The GEOSS also improves the degree of accuracy with which information is provided. Although many studies assume more information is valuable, in reality this is not always the case. A study on prevention of potentially harmful algae blooms in the North Sea indicated that an early warning system with a considerable probability of a false warning (type-II error of some 20%) has no value (Bouma et al. 2009).

Managing the interlinked challenges of global change will require an increasing flow of information about developments in the global environment and the world's societies that interact with it. International agreements, as well as national management strategies, will be on weak footing unless relevant data streams are established with the foresight to support decision-making and the science that underpins it. With national contributions to GEOSS—a global public information good—now increasingly under pressure due to overall budget constraints and investments in observation infrastructure not keeping pace with the foreseeable demand for information, demonstrations of the potential benefits to global societies as well as of the impacts of failed international cooperation are a matter of increasing urgency. This is because implementation of global Earth observation assets and Earth system models require significant time from the development of technology and methods to their operational deployment.

Our assessment found that in the majority of case studies, the societal benefits of improved and globally coordinated Earth observation systems were orders of magnitude higher than the investment costs. A strong coordinating institution is required to ensure that an integrated architecture takes full advantage of the increased benefits and cost reductions achieved by international cooperation. Furthermore, boundary organizations interfacing between the science community and users such as businesses, and governmental and nongovernmental organizations have to be an integral part of an enhanced observation strategy in support of global change management.

Continuous and comprehensive monitoring of the Earth carries the potential for major advancements in global change science. Not only will science generate more robust knowledge through data assimilation into increasingly integrated Earth system models, new scientific fields will also arise. Pairing environmental monitoring with next-generation acquisition of socioeconomic parameters will yield new insights into possible societal pathways through the problematic bottleneck of twenty-first-century environmental exploitation on the road to increased sustainability. New GEOSS-informed Earth system science products and services will emerge and will need to be assisted to diffuse for wider societal use. The benefits from these emerging applications are among the least predictable today, but they must receive adequate support to guarantee a transition to more science- and evidence-based decision-making.

References

Böttcher H et al (2009) An assessment of monitoring requirements and costs of Reduced Emissions from Deforestation and Degradation'. Carbon Balance Manage 4:7

Bouma JA et al (2009) Assessing the value of information for water quality management in the North Sea. J Env Man 90(2):1280–1288

Fritz S et al (2009) Geo-wiki.org: the use of crowdsourcing to improve global land cover. Remote Sens 1(3):345–354

GEOBENE (2016) About the GEOBENE project| GEOBENE project. http://www.geo-bene.eu. Accessed 7 Nov 2016

Rao S (2009) Scaling up low carbon investments: focus on renewable energy in Latin America. http://pure.iiasa.ac.at/9029/. Accessed 8 Dec 2016

Searchinger TD et al (2009) Fixing a critical climate accounting error. Science 326:527–528

Walker B et al (2009) Looming global-scale failures and missing institutions. Sci Environ 325:1345–1346

Author Biography

Michael Obersteiner is program director of the Ecosystems Services and Management (ESM) Program at the International Institute for Applied Systems Analysis (IIASA) in Laxenburg, Austria. He joined IIASA's Forestry Program (FOR) in 1993 and has been leading the Group on Global Land-Use Modeling and Environmental Economics since 2001. His background includes the fields of global terrestrial ecosystems and economics, specializing in REDD and REDD+ modeling as well as policy assessments, with particular expertise on the tropical forest zones of South America, Africa, and Asia.

Dr. Obersteiner completed graduate studies both in Austria (BOKU University and the Institute for Advanced Studies Vienna) and abroad (Columbia University, New York, and the Siberian Branch of the Russian Academy of Sciences, Novosibirsk). Since 2004, he has made substantial contributions to the development, establishment, and management of the IIASA-ESM integrated modeling cluster, which includes widely recognized global biophysical and economic models in the area of agriculture, forestry, and land use (G4M, EPIC, GLOBIOM). In 2008–2009, Dr. Obersteiner served as a seconded staff expert for the Group on Earth Observations (GEO) in Geneva,

Switzerland. Over the past decade, Dr. Obersteiner has been the principal investigator at IIASA for more than 30 international projects covering diverse fields of different scales and numerous funding organizations. Dr. Obersteiner has been a consultant to a number of national and international organizations, including inter alia the European Commission, WWF, OECD, and other national and international institutions. He has authored more than 250 scientific papers and consultancy reports in the aforementioned fields.

Chapter 13
Earth Observation Support to the UN Framework Convention on Climate Change: The Example of REDD+

Ake Rosenqvist

13.1 Background

Forests present a significant global carbon stock accumulated primarily through growth of trees and an increase in soil carbon. While sustainable management, planting, and rehabilitation of forests can conserve or increase forest carbon stocks, deforestation, degradation, and poor forest management result in reductions. According to estimates from the Global Forest Resources Assessments (FRA) of the Food and Agriculture Organization of the United Nations (FAO) (FAO 2016), global forest biomass stocks decreased at an estimated average rate of 2.5% per year during the period 1991–2015, with about 2% resulting from deforestation and 0.5% from forest degradation (Federici et al. 2015).

The large contribution of greenhouse gas emissions from deforestation and forest degradation—in particular from non-Annex-I (developing) countries—is recognized in international climate negotiations. In 2005 the term "REDD", standing for "Reducing emissions from deforestation and forest degradation" was introduced to the UNFCCC following an initiative by the governments of Papua New Guinea and Costa Rica (UNFCCC 2005). During subsequent refinement of the concept at UNFCCC COP meetings (Bali, Poznań, Copenhagen, Cancun, Warsaw, Paris), it became commonly known as "REDD+" because of a broadening of scope. Since the Cancun COP, REDD+ comprises five specific activities for which results-based incentives for participating countries can be made:

- Reducing emissions from deforestation;
- Reducing emissions from forest degradation;
- Conservation of forest carbon stocks;
- Sustainable management of forests;

A. Rosenqvist (✉)
solo Earth Observation (soloEO), Tokyo, Japan
e-mail: ake.rosenqvist@soloEO.com

© The Author(s) 2017
M. Onoda and O.R. Young (eds.), *Satellite Earth Observations and Their Impact on Society and Policy*, DOI 10.1007/978-981-10-3713-9_13

- Enhancement of forest carbon stocks.

Countries participating in REDD+ (on a voluntary basis) are required to report their national estimates of carbon dioxide and other greenhouse gas emissions and removals resulting from significant REDD+ activities to the UNFCCC on a biennial basis, following an agreed transparency framework for Measurement, Reporting, and Verification (MRV) (United Nations Climate Change Secretariat 2014). They are requested to utilize Intergovernmental Panel on Climate Change (IPCC) guidance and guidelines and to establish National Forest Monitoring Systems (NFMS) to estimate anthropogenic emissions and removals, using a combination of remote sensing and ground-based data.

The potential of remote sensing technology to support UNFCCC was first assessed more than a decade and a half ago (Rosenqvist et al. 2003), recognizing the significant potential of satellite-based techniques to monitor the state and changes of forest and land cover over extensive areas. Operational uptake of the technology has been slower than anticipated, with relatively few developing countries using satellite data as a tool for national forest monitoring in the first decade following the assessment (the most prominent exception being Brazil). The inertia of major space agencies and satellite data operators in recognizing and responding to the opportunity was probably at least part of the reason for the slow progress. Restricted/commercial data policies in particular have long been a key obstacle to widespread operational uptake of the technology. The US decision in 2008 to release all Landsat satellite data for unrestricted use was a major game-changer for Earth observation, prompting other countries to realize the advantages of free and open data access.

Many developing countries are now actively engaged (through REDD+) and are making use of ever-increasing remotely sensed datasets thanks to global systematic acquisition strategies and free and open access to a number of key Earth observation satellites. Such data assist in defining the extent of human activity causing emissions and removals from forest lands including historical impacts in the development of Forest Reference [Emission] Levels (FRELs/FRLs).[1]

13.2   GFOI Goals and Objectives

The Global Forest Observations Initiative (GFOI) has been established to facilitate the supply and use of Earth observation data to countries as part of developing national forest monitoring systems for the purposes of REDD+. Through collaborative action, the initiative seeks to foster forest monitoring and assessment that is

---

[1]Benchmarks are expressed in tonnes CO2 equivalent per year for assessing each country's performance in implementing REDD+ activities.

robust, reliable, and achievable at reasonable cost and supports planning for national development priorities including climate change mitigation and adaptation.

GFOI was founded under the intergovernmental Group on Earth Observations (GEO). First initiated as a demonstration activity—the Forest Carbon Tracking (FCT) Task—in 2008, the first GFOI Implementation Plan was approved by the GEO Plenary in 2011, followed by long-term governance arrangements established in 2013. The GFOI is led by a Lead Team consisting of FAO, CEOS, Australia,[2] Norway[3] and the U.S.,[4] with support from other governments and institutions. It benefits from advisory committees with representatives from the UNFCCC, the IPCC GHG inventory program, the World Bank Forest Carbon Partnership Facility (FCPF), Global Observation for Forest Cover and Land Dynamics (GOFC-GOLD), and various other institutions and experts.

GFOI sets out to facilitate widespread implementation of REDD+. It aims to support countries that are developing and implementing NFMS and associated emissions MRV systems to take full advantage of Earth observation technology, by encouraging the use and assuring sustained availability of satellite observations required for forest monitoring; engaging space agencies and satellite data providers; developing methods and protocols on the combined utilization of remotely sensed and ground data for transparent assessment and reporting; developing capacity building programs to provide sustained assistance and guidance; and promoting directed research and development on high priority topics where further development is needed (e.g. forest degradation, above-ground biomass, satellite data synergy).

Reporting greenhouse gas emissions and removals to the UNFCCC is the responsibility of the countries that voluntarily opt to participate in REDD+. GFOI aims to *support* countries developing their national forest monitoring systems, not to conduct monitoring on their behalf. Countries are engaged on equal terms and hold full ownership over their NFMS and associated capacities.

The sections below provide a brief summary of some early achievements and challenges of the four components of GFOI: Satellite Data Coordination; Methods and Guidance; Capacity Building; and Research & Development Coordination.

---

[2]Australian Department of the Environment.

[3]Norwegian International Climate and Forest Initiative (NICFI), Ministry of Climate and Environment (Klima- og miljødepartementet).

[4]SilvaCarbon—technical cooperation program of the U.S. Agency for International Development (USAID), the U.S. Forest Service within the Department of Agriculture (USFS), the U.S. Geological Survey of the Department of Interior (USGS), the U.S. Environmental Protection Agency (EPA), the U.S. Department of State, the National Aeronautics and Space Administration (NASA), the National Oceanic and Atmospheric Administration within the Department of Commerce (NOAA), and the Smithsonian Institution.

13.3   Achievements and Challenges

13.3.1   Data Coordination

The Committee on Earth Observation Satellites (CEOS) coordinates civil space-borne observations of the Earth and leads the GFOI Data Coordination component. A dedicated Space Data Coordination Group for GFOI (SDCG) was established by CEOS and charged with developing and coordinating the satellite data acquisition strategies required in support of countries participating in REDD+. A key strength of SDCG is the broad support and active participation it enjoys from the national space agencies and data providers that are operating the key satellite missions of relevance for REDD+ (Table 13.1).

Restricted data policies and inconsistent data archives have long been major obstacles to operational utilization of satellite data over national scales (Reiche et al. 2016; Rosenqvist et al. 2003). In order to avoid creating a dependency on commercial satellite data for reporting, SDCG primarily focuses, as a matter of principle, on the coordination of freely available, sharable, and accessible global data. Data from the US Landsat series is currently most commonly used by countries, with data from the European Sentinel satellite constellation—also publicly open— expected to grow in importance as the satellites (first unit launched in 2014) and ground segment become fully operational. As an indication of the significance of these open data policy missions, SDCG is jointly chaired by USGS and ESA.

Table 13.1 Participating space data providers and corresponding satellite missions (as of May 2016)

Satellite mission	Type	Data policy	Agency/Data provider	Country
Landsat 7/8	Optical	Public open	USGS (NASA)	U.S.
Sentinel-1	Radar	Public open	ESA/European commission	Europe/EU
Sentinel-2	Optical	Public open	ESA/European commission	Europe/EU
CBERS-4	Optical	Public open*	INPE and CRESDA	Brazil and China
COSMO-SkyMed	Radar	PPP**	ASI	Italy
Pléiades	Optical VHR	PPP**	CNES	France
Radarsat-2	Radar	PPP**	CSA	Canada
TerraSAR-X/TanDEM-X	Radar	PPP**	DLR	Germany
ALOS-2	Radar	PPP**	JAXA	Japan
SPOT-6/7	Optical	Commercial	Airbus Defence & Space	Europe

* CBERS—public open from INPE, ** *PPP* Public-Private Partnership, *VHR* Very-High Resolution

Agencies with semi-commercial or Public–Private Partnership (PPP) missions also contribute to GFOI by making data available to the GFOI R&D program (see Sect. 13.3.4). As all aspects of national forest monitoring cannot be fully addressed solely with Landsat and Sentinel-type data, including priority REDD+ topics still in the research realm, the additional information provided by the very-high resolution and long-wavelength radar sensors operated by these agencies can often be part of the solution.

SDCG is not primarily about data provision per se. One of its main functions is to support CEOS with strategic solutions to better meet the requirements of REDD+ reporting. REDD+ countries commonly face different challenges compared to other users that CEOS agencies traditionally aim to support. These challenges include poor Internet connectivity, limited data processing capacity, and incomplete data archives, made more challenging by the possible requirement to handle large volumes of data covering national territories at a high temporal frequency.

SDCG actively promotes the implementation of systematic satellite data acquisition strategies to CEOS agencies to assure that *all* forested regions on the Earth are covered with sufficient temporal frequency and with as many different sensor types as possible.

To address Internet infrastructure issues and reduce data complexity, SDCG is, in collaboration with FAO and a number of pilot countries, developing cloud-based processing and analysis solutions and storage of preprocessed ready-to-use satellite data (also referred to as Analysis-Ready Data, ARD) in seamless "Data Cubes". It is hoped that novel concepts such as these will contribute to reducing the current obstacles to the uptake and application of satellite data in REDD+ countries and increase their ownership of the process, and they can also be expected to provide innovative solutions on how space agencies can handle the increasing amount of satellite data more efficiently in the future.

13.3.2 Methods & Guidance Documentation

The GFOI Methods & Guidance Documentation (MGD)[5] provides methodological advice on the joint use of remotely sensed and ground-based data to estimate and report greenhouse gas emissions and removals associated with forests in a manner consistent with the greenhouse gas inventory guidance from the IPCC. This is required by decisions of the UNFCCC for voluntary implementation of REDD+ activities.[6]

---

[5]Available in English, French and Spanish at http://www.gfoi.org/methods-guidance/.

[6]The REDD+ activities as listed in the Cancun Agreements (UNFCCC Decision 1/CP.16 para 70) are: (a) Reducing emissions from deforestation; (b) Reducing emissions from forest degradation; (c) Conservation of forest carbon stocks; (d) Sustainable management of forests; (e) Enhancement of forest carbon stocks.

The document is written by a team of internationally renowned experts and stakeholders specializing in forest information systems and the observations underpinning those systems, as well as experts in the assessment of forest carbon emissions and removals. The MGD is complementary to the IPCC Good Practice Guidance (GPG) and the GOFC-GOLD REDD Sourcebook (GOFC-GOLD 2015).

The second edition of the MGD, entitled *Integrating remote sensing and ground-based observations for estimation of emissions and removals of greenhouse gases in forests: Methods and Guidance (MGD) from the Global Forest Observations Initiative*, was released in 2016. The MGD is available for free from the GFOI website.

The MGD has been well received by both participating REDD+ countries and key stakeholders and has contributed to:

- Building a bridge between the decisions of the UNFCCC and operational decisions made by countries opting to estimate GHG emissions and removals from REDD+ activities;
- Systematically describing how IPCC Emissions Inventory methods can be used to make REDD+ estimates, activity by activity, using a combination of remotely sensed and ground-based data;
- Defining a foundation for consistent methodological advice, training, technical support, and communication provided to countries through GFOI engagement;
- Providing a link to freely available satellite data sets (via the GFOI Space Data Coordination Component); and,
- Communicating REDD+ country methodological requirements and data needs to the research and development community.

A key to success for the MGD has been the close linkage (facilitated by GFOI) to the relevant policy, implementation, funding, methodology, and research organizations involved in the REDD+ process (e.g. UNFCCC, IPCC, World Bank, SilvaCarbon, NICFI, UN-REDD, and GOFC-GOLD). Interactions between these key REDD+ support agencies combined with government partners working bilaterally on REDD+ initiatives have enabled the MGD to respond to REDD+ countries' methods and guidance needs.

Additionally, feedback from users on the utility of the MGD (received through FAO and SilvaCarbon capacity building workshops) has led to a number of innovations designed to address the challenges of effectively communicating technical guidance in a dynamic and rapidly emerging sector to a multilingual global audience with specific national needs.

These innovations include:

- Publication of rapid response guidance modules that expand on guidance presented in MGD in topical areas such as forest reference emission levels and the use of global data sets; and

- Release of REDDcompass, a web application providing access to the MGD in an interactive online framework (GFOI 2016). REDDcompass enables users to progressively work through the key themes, concepts, and actions of REDD+ NFMS for MRV, gaining access to a suite of GFOI methods and guidance, space data resources, training materials, and tools along the way.

The MGD builds on the foundation of the first edition and enhances and extends guidance on institutional arrangements, the use of global data sets, forest reference emission levels, safeguards, integration methods, extended systematic statistical advice on emission and removal factors and activity data, and reporting and verification requirements and processes. All methods and guidance materials and supporting references and tools are available through REDD compass.

13.3.3   Capacity Building

The GFOI Capacity Building component provides the interface between GFOI and the REDD+ countries that are its target users. The US SilvaCarbon program manages the Capacity Building component of GFOI in collaboration with FAO/UN-REDD.

SilvaCarbon implements capacity building efforts at both national and regional levels. At national levels, it works with countries to develop country-specific work plans ranging from technical assistance to intensive technical training sessions, workshops, and study tours. They are provided by experts from the U.S. SilvaCarbon institutions, and often in collaboration with other GFOI organizations and affiliated partners from industry and academia. The regional work plans are developed to reach out to a larger number of countries that share some common attributes or interests (e.g. in Central America, the Andean Amazon region, Central Africa and South/Southeast Asia). They typically include activities such as hands-on or showcase workshops on a range of topics related to monitoring and managing forest and terrestrial carbon in the regions.

FAO, through UN-REDD, supports national REDD+ readiness efforts in partner countries through direct support to the design and implementation of UN-REDD National Programs, as well as through complementary support to national REDD+ actions through common approaches, analyses, methodologies, tools, data, and best practices. Both SilvaCarbon and FAO have an active presence in partner countries and are well positioned to serve as links between countries and the other components of GFOI as they relate to capacity building needs. Both use and promote the GFOI MGD in training events and in turn provide comments and feedback from country representatives to the MGD team. There is also close cooperation with the Data Coordination component, to report user needs to CEOS space agencies and data providers, as well as to prepare and provide satellite data for training and NFMS development.

Long-term engagement is a prerequisite to building and maintaining technical capacity in a region. Workshops are never one-time events, but undertaken on a regular basis over several years and on a range of specialized topics. In the Americas region for example, SilvaCarbon events often engage the same core group of participants from the relevant organizations in the countries, which gradually build up sustained capacity on different aspects of the REDD+ process. Importantly, the events also stimulate personal and institutional networking between participants and organizations in the region. Regional collaboration and support is of mutual benefit for all parties.

There are naturally huge challenges in building up capacity in a large number of countries in a relatively short period of time and with finite staffing and financial resources. Sustained in-country engagement is in most cases not feasible. The level of basic technical infrastructure and competence—including familiarity with Earth observation techniques—can also vary greatly between countries and regions, posing an additional difficulty to regional capacity building efforts. Language is still a complication, despite publication of the MGD in three languages. While workshops in the Americas region are commonly undertaken partly or fully in Spanish, most training events in Africa and Asia are undertaken in English.

From an organizational point of view, both SilvaCarbon and FAO/UN-REDD are independent undertakings, with objectives, timelines, and budgets formally separate from GFOI. Coordination between SilvaCarbon, FAO, and the different GFOI components facilitates the success of the GFOI Capacity Building component in helping increase the use and awareness of Earth observation techniques for the REDD+ process. In 2016, GOFC-GOLD (with the support of the Forest Carbon Partnership Facility) started coordinating training workshops with SilvaCarbon and the FAO. Such joint efforts improves the cost-efficiency of the activities.

13.3.4 Research and Development Coordination

The GFOI R&D Coordination (RDC) component is led by the GOFC-GOLD Land Cover Office, with support from ESA. The RDC component covers focused R&D actions addressing immediate needs for NFMS, rather than more long-term, basic research activities in the forest monitoring domain. The primary objective of the RDC component is to support the MGD component by advancing select topics of priority for REDD+ implementation to reach a technical readiness level sufficient for inclusion in the MGD. The RDC component also supports the development of training materials. Key research needs have been prioritized in a peer-reviewed report (GFOI 2013), addressing priority R&D topics such as methods for mapping forest degradation and particular forest types (e.g. mangrove, peat forests, woodlands), vegetation biomass estimation, and satellite sensor synergy.

To capitalize on the significant forestry-related research that is already underway worldwide, the RDC component is engaged in two main activities: the organization of dedicated Expert workshops and the management of an R&D program for GFOI.

The Expert workshops are typically focused on one or two selected priority topics and seek to identify the main obstacles to their widespread/operational implementation. In case the challenges are related to satellite data (e.g. lack of adequate time series, cost of data, or sensor shortcomings), comments and recommendations are channeled back to CEOS and SDCG. For challenges of a methodological nature, the workshops provide opportunities to pool capacity and brainstorm alternative solutions. SDCG coordinates access to a wide range of public and commercial test data for high value research areas to support experts in the development of new solutions.

The RDC is set up to engage external research groups and help focus their activities towards the GFOI Priority R&D Topics and dedicated country needs. It is supported by the SDCG space agencies, which provide dedicated satellite data—also from commercial and PPP missions—to progress priority research and development of related novel Earth observation techniques.

The main challenge to GFOI R&D is financial. GFOI funds are limited to management and coordination only, with no budget allocation to support R&D per se. This leads to reliance on external R&D efforts, which may only partly overlap with the goals and timelines of the GFOI. Nevertheless, recent research calls (e.g. Innovators from ESA) have been based on the recommendations from the RDC component. One can anticipate more research calls will be based on the GFOI activities in the future. The RDC setup with Expert workshops and a "data-for-research" R&D program is a solution that nonetheless has proven valuable.

13.4   Conclusions

The GFOI constitutes an example of how Earth observation data can be used to support Multilateral Environmental Agreements (MEA), such as the UNFCCC, or the new Sustainable Development Goals (SDGs). While it is too early to assess the full impact of GFOI on REDD+ implementation, the initiative is certainly in a good position with active participation by developing countries and multilateral organizations, including those providing funding. With the backing and active engagement of CEOS space agencies, key organizations and stakeholders engaged in REDD+, national governments, and technical and scientific experts on remote sensing and forestry, circumstances can hardly get better. The stars are aligned. It is now or never to prove the worth of Earth observation.

Acknowledgements The views expressed are those of the author who is grateful to the late Jim Penman, Carly Green, Sylvia Wilson, Frank Martin Seifert, Eugene Fosnight, Thomas Harvey, and Brice Mora for reviewing the manuscript.

References

FAO (2016) Global forest resources assessments. http://www.fao.org/forest-resources-assessment/current-assessment/en/. Accessed 12 Aug 2016

Federici S et al (2015) New estimates of CO_2 forest emissions and removals: 1990–2015. For Ecol Manage 352:89–98

GFOI (2013) Review of priority research & development topics: R&D related to the use of remote sensing in national forest monitoring. http://www.gfoi.org/wp-content/uploads/2015/03/GFOI_ReviewPrioityRDTopics_V1.pdf. Accessed 7 Dec 2016

GFOI (2016) GFOI REDDcompass. https://www.reddcompass.org/frontpage. Accessed 12 Aug 2016

GOFC-GOLD (2015) GOFC-GOLD REDD sourcebook. http://www.gofcgold.wur.nl/redd/sourcebook/GOFC-GOLD_Sourcebook.pdf. Accessed 12 Aug 2016

Reiche J et al (2016) Combining satellite data for better tropical forest monitoring. Nature Climate Change 6:120–122

Rosenqvist A et al (2003) A review of remote sensing technology in support of the Kyoto Protocol. Environ Sci Policy 6(5):441–455

UNFCCC (2005) Reducing emissions from deforestation in developing countries: approaches to stimulate action—submissions from Parties. http://unfccc.int/resource/docs/2005/cop11/eng/misc01.pdf. Accessed 12 Aug 2016

United Nations Climate Change Secretariat (2014) Handbook on measurement, reporting and verification for developing country Parties. https://unfccc.int/files/national_reports/annex_i_natcom_/application/pdf/non-annex_i_mrv_handbook.pdf. Accessed 12 Aug 2016

Author Biography

Ake Rosenqvist is CEO of solo Earth Observation (soloEO), an SME providing scientific consulting in the field of satellite-borne environmental monitoring and mission planning. He joined the Swedish Space Corporation (SSC) in 1990 and was the first foreign national invited to join the National Space Development Agency of Japan (NASDA) in 1993, where he became involved in the development of the JERS-1 applications program. He spent 6 years at the EC Joint Research Centre in Ispra, Italy, and was awarded the ISPRS President's Honorary Citation in 2000 for promotion of Earth observation applications to the United Nations Framework Convention on Climate Change (UNFCCC) Kyoto Protocol. As senior scientist at the Japan Aerospace Exploration Agency (JAXA), he developed the concept of systematic data observations, which was adopted by JAXA and implemented for the ALOS and ALOS-2 missions. Dr. Rosenqvist founded soloEO in 2009 and is currently supporting JAXA, the Argentinian Space Commission (CONAE), and the

European Space Agency (ESA) with planning for the ALOS-2, SAOCOM-1, and SAO-CS radar missions. On behalf of JAXA and the Norwegian Space Centre (NSC), soloEO has supported the development of the Global Forest Observations Initiative (GFOI) since its establishment in 2008.

Chapter 14
Quantitative Assessment of the Earth Observation Data and Methods Used to Generate Reference Emission Levels for REDD+

Brian Alan Johnson, Henry Scheyvens and Hiromitsu Samejima

14.1 Introduction

14.1.1 REDD+ Forest Reference Emission Levels and Baselines

At the 21st Conference of the Parties (COP) in 2015, Parties to the UNFCCC set out an ambitious plan—the Paris Agreement—for global action on climate change mitigation and adaption. To mitigate climate change, the Paris Agreement promotes the implementation of policy approaches and positive incentives for reducing emissions from (a) deforestation and (b) forest degradation, and for increasing (c) the conservation of forest carbon stocks, (d) the sustainable management of forests, and (e) the enhancement of forest carbon stocks (or what is commonly referred to as REDD+). Two years prior, at the 19th COP in 2013, Parties agreed on a set of decisions known as the Warsaw Framework for REDD+, which among other things, set guidelines for the development of national (or in the interim, subnational) forest monitoring systems (UNFCCC 2013a) and forest reference emission levels/forest reference levels (FRELs) (UNFCCC 2013b), which are required for countries to receive results-based payments for REDD+ activities (UNFCCC 2010).

B.A. Johnson (✉) · H. Scheyvens · H. Samejima
Natural Resources and Ecosystem Services, Institute for Global Environmental Strategies,
2108-11 Kamiyamaguchi, Hayama, 240-0115 Kanagawa, Japan
e-mail: johnson@iges.or.jp

H. Scheyvens
e-mail: scheyvens@iges.or.jp

H. Samejima
e-mail: samejima@iges.or.jp

© The Author(s) 2017
M. Onoda and O.R. Young (eds.), *Satellite Earth Observations and Their Impact on Society and Policy*, DOI 10.1007/978-981-10-3713-9_14

155

A FREL is a projected level of greenhouse gas emissions (or removals, which are considered negative emissions), given in tons per year for a specific timeframe in the future. It is calculated based on historical emissions from one or more of the REDD+ activities of (a–e) above, with adjustments based on national circumstances permitted when justified. The FREL serves as a baseline to measure a country's performance in mitigating climate change in the forest sector. Specific methodologies for calculating historical emissions and generating FRELs are not stipulated in UNFCCC decisions, which provides flexibility by allowing for differences in national circumstances/capacities and for FRELs to be improved over time as national capacities and data quality/quantity increase (UNFCCC 2011). However, at COP15, Parties agreed that the historical emissions must be calculated using a combination of remote sensing (RS) and ground-based forest carbon inventory (GBFCI) approaches (UNFCCC 2009). The Earth observation data are typically processed using RS image-processing techniques (e.g., supervised classification and change detection algorithms) to map forest changes over time, while the resultant emissions from these changes are estimated using the GBFCI data (Intergovernmental Panel on Climate Change 2006). After calculating the historical emissions, future emissions are then projected as the final step of generating the FREL. Various modeling approaches exist for projecting future emissions and some common approaches include: the use of the historical average emissions rate (i.e., assuming no change in future emissions), temporal regression modeling using historical emissions data (with or without other nonspatial data that reflect the drivers of historical emissions), and spatiotemporal modeling using historical emissions data (with or without other nonspatial data) as well as spatial data that reflect the drivers of historical emissions (e.g., slope maps, locations of roads, etc.) (Busch et al. 2009; Huettner et al. 2009).

Apart from its use for generating national/sub-national FRELs, Earth observation data are also being used for REDD+ implementation at the individual site or project level, and many REDD+ projects are ongoing or already completed in voluntary carbon trading markets. The development of REDD+ projects for these voluntary carbon trading markets was spurred by a decision from the 13th COP in 2007, which encouraged REDD+ demonstration activities and invited developed countries to mobilize resources to support these efforts (UNFCCC 2007). To assure the quality of REDD+ projects, various certification schemes have been established, with the Verified Carbon Standard (VCS) being one of the largest certifiers of projects selling credits into the forest carbon markets (Hamrick and Goldstein 2015; McDermott et al. 2012). In contrast to the FRELs, for VCS-certified REDD+ projects there are detailed methodological requirements on how the future emissions levels (hereafter referred to as "baselines" to differentiate them from FRELs) can be projected. For example, VCS methodologies for generating baselines of emissions from unplanned future deforestation and/or forest degradation (DD) require that the future rate of DD in the project area be estimated based on the historic rates of DD in an area where the agents and drivers of deforestation as well as the soil type/slope/elevation are similar to those found in the project area (e.g., in a larger "reference area") (Verified Carbon Standard 2013), so Earth observation data of the "reference area" are needed to generate the baseline.

From this discussion, it is clear that Earth observation data are essential for developing FRELs and baselines at the national/subnational and project levels, respectively. Thus, the Earth observation derived historical emissions estimates and the projections of future emissions must be accurate to ensure the contributions of REDD+ activities to climate change mitigation are correctly quantified. The Earth observation data quality (e.g., spatial resolution of the data) has a direct impact on the accuracy of the historical forest maps used to estimate emissions (Hilker et al. 2009), while the accuracy of some modeling approaches for projecting future emissions (e.g., regression models) will also be affected by the Earth observation data quantity (number of historical forest maps produced). For these reasons, it is necessary to assess the quality and quantity of Earth observation data being used, as well as how the data are being used, to really understand the contribution of Earth observation to the implementation of REDD+.

14.1.2   Objectives

This study had two main objectives. The first was to better understand the quality and quantity of Earth observation data being used for REDD+ at the national/subnational and project levels. New sources of Earth observation data are becoming available every year, so it should be possible to increase the quality and quantity of the data used for REDD+ FRELs/baselines over time. To evaluate the quality and quantity of the Earth observation data currently being used at the country and project levels, we computed four metrics related to the spatial and temporal resolution and scale of the data. A previous study (Johnson et al. 2016) performed a similar analysis based on the Earth observation data used by the six countries that submitted their proposed FRELs by the end of 2015, but in this chapter we expand the scope of analysis to include nine additional countries (those that submitted their proposed FRELs in 2016) as well as 15 REDD+ projects.

The second objective was to understand whether or not the Earth observation data quality and quantity were determining the types of REDD+ activities and modeling approaches used to generate FRELs/baselines. For this, we computed two additional metrics related to methods used for FREL/baseline generation: the number of REDD+ activities assessed by Earth observation data for FREL/baseline generation and the level of complexity of the model used to project future emissions. The types of activities that Earth observation data can be used to accurately estimate historical emissions for, as well as the level of complexity of the model used to project future emissions, are both affected by the Earth observation data quality and quantity, so we hypothesized that countries/projects using higher quality and quantity Earth observation data should be assessing historical emissions from more REDD+ activities and also using more complex models to project future emissions. If this is not the case, some factors other than the Earth observation data quality and quantity may be the primary determinants of the REDD+ activities assessed and/or the models selected to project future emissions.

14.2   Methods

14.2.1   Data

As of April 2016, 15 countries have submitted reports for technical review to the UNFCCC secretariat containing the details of their proposed FRELs.[1] The reports contain detailed information on the Earth observation data that countries used to estimate historical emissions as well as the approaches they used to project future emissions from the historical data. This information was used to calculate the six metrics in this study. At the project level, similar information was obtained for a sample of 15 VCS-verified REDD+ projects (Verified Carbon Standard 2016a). Tables 14.1 and 14.2 list the countries that have submitted their proposed FRELs to the UNFCCC as of April 2016, as well as the locations of the 15 sample VCS-verified REDD+ projects.

14.2.2   Metrics

Four metrics were calculated to assess the quality and quantity of Earth observation data used to estimate historical emissions:

$%mapped$ percentage of country (or reference area for projects) mapped using Earth observation data

$$(14.1)$$

$%mapped$ length of historical assessment period for which Earth observation data was used, in years (14.2)

$#maps$ the number of maps produced during the historical assessment period (14.3)

$spatial\_res$ spatial resolution of Earth observation data (14.4)

---

[1]Available online at: http://redd.unfccc.int/fact-sheets/forest-reference-emission-levels.html.

Table 14.1 Metric values calculated at the national/subnational level

Country	% mapped	timeframe	#maps	spatial_res	#activities	model_complexity
Brazil	49%	14	15	30 m	1 (DF)	1
Chile	22%	Varied by region		30 m	4 (DF, FD, FCSE, CFCS)	1
Colombia	40%	12	7	30 m	1 (DF)	1
Congo	100%	12	2	30 m	2 (DF, FD)	1
Costa Rica	95%	12	3	30 m	2 (DF, FCSE)	1
Ecuador	100%	8	2	30 m	1 (DF)	1
Ethiopia	100%	13	2	30 m	2 (DF, FCSE)	1
Guyana	100%	11	7	30 m	2 (DF, FD)	1
Indonesia	60%	22	8	30 m	2 (DF, FD)	1
Malaysia	100%	13	3	30 m	1 (SFM)	1
Mexico	100%	10	3	30 m	1 (DF)	1
Paraguay	100%	15	5	30 m	1 (DF)	1
Peru	61%	13	14	30 m	1 (DF)	2
Vietnam	100%	15	4	30 m	5 (DF, FD, FCSE, CFCS, SFM)	1
Zambia	100%	14	3	30 m	1 (DF)	1
Median (range)	100% (22–100)	13 (8–22)	3.5 (2–15)	30 m (30)	1 (1–5)	1 (1–2)

DF deforestation, *FD* forest degradation, *FCSE* forest carbon stock enhancement, *CFCS* conservation of forest carbon stocks, *SFM* sustainable forest management

Of these metrics, Eqs. 14.1–14.3 are indicators of the Earth observation data quantity (area extent and temporal scale/resolution), while Eq. 14.4 is an indicator of the data quality because it determines the minimum changes in forest extent that can be detected. Because the value of Eq. 14.2 may have an effect on the value of Eq. 14.3 (a longer *timeframe* allows for a greater *#maps*), we assessed the correlation between the two metrics using the Spearman rank correlation test ($p < 0.05$) (Wayne 1990).

For countries/projects that generated multiple FRELs/baselines using different historical assessment periods, the value reported in Eq. 14.2 is that of the most recent historical assessment period. Additionally, some countries/projects used forest maps from before or after the designated historical assessment period to estimate the historical emissions (i.e., by interpolating the forest extent from maps

Table 14.2 Metric values calculated at the project level

Project name	Applied methodology	% mapped	timeframe	#maps	spatial_res	#activities	model_complexity
ADPML Portel-Pará REDD Project—Brazil	VM0015	100%	13	3	30 m	1 (DF)	1
Alto Mayo Conservation Initiative	VM0015	100%	11	3	30 m	1 (DF)	2
Biocorridor Martin Sagrado REDD+ Project	VM0015	100%	11	3	30 m	1 (DF)	1
Carbon Emissions Reduction Project in the Forest Corridor Ambositra-Vondrozo	VM0015	100%	16	3	30 m	1 (DF)	1
Chocó-Darién Conservation Corridor REDD Project	VM0009	100%	25	7	30 m	1 (DF)	2
Cordillera Azul National Park REDD Project	VM0007	100%	15	3	30 m	1 (DF)	2
Florestal Santa Maria Project	VM0007	100%	12	5	30 m	2 (DF, FD)	1
Isangi REDD+ Project	VM0006	100%	16	3	30 m	1 (DF)	3
Kariba REDD+ Project	VM0009	100%	9	6	30 m	2 (DF, FD)	2
Kasigau Corridor REDD Project Phase I Rukinga Sanctuary	VM0009	100%	18	7	30 m	1 (DF)	2
Kulera Landscape REDD+ Program for CoManaged Protected Areas	VM0006	100%	12	3	30 m	1 (DF)	3
Lower Zambezi REDD+ Project	VM0009	100%	26	6	30 m	1 (DF)	2
Madre de Dios Amazon REDD Project—FSC concessions	VM0007	100%	9	3	30 m	1 (DF)	2
Purus Project	VM0007	100%	11	12	30 m	1 (DF)	1
REDD in Community Forests—Oddar Meanchey	VM0006	100%	19	6	30 m	2 (DF, FCSE)	1
Median (range)		100% (100)	13 (9–26)	3 (3–12)	30 m (30)	1 (1–2)	2 (1–3)

DF deforestation, *FD* forest degradation, *FCSE* forest carbon stock enhancement, *CFCS* conservation of forest carbon stocks, *SFM* sustainable forest management generated before/after the assessment period), so in these cases the values for both Eqs. 14.2 and 14.3 were changed to match the first and last image years to better reflect the actual Earth observation data used. Finally, for countries/projects that used multiple Earth observation data sources (e.g., because finer spatial resolution data became available in more recent years), the value reported for Eq. 14.4 was that of the coarsest resolution data set because this data set limits the finest-scale changes that can be detected over the entire historical assessment period. After calculating these metrics, median values were computed to show the typical quality and quantity of Earth observation data being used at the national/subnational and project levels.

In addition to the four Earth observation data quality/quantity metrics, two additional metrics were calculated to assess how the Earth observation derived historical emissions were used for FREL/baseline generation:

$$\#activities \quad \begin{array}{l} \text{number of REDD + activities assessed by Earth observation} \\ \text{data for FREL/baseline generation} \end{array} \quad (14.5)$$

$$model\_complexity \quad \begin{array}{l} \text{level of complexity of the model used to project future} \\ \text{emissions} \end{array} \quad (14.6)$$

Some countries/projects assessed historical emissions for multiple REDD+ activities using Earth observation data, but did not include all of them in their FREL/baseline due to an unacceptable level of estimation uncertainty, so the value reported for Eq. 14.5 was limited to the number of REDD+ activities actually included in the FREL/baseline. For Eq. 14.6, we assigned ordinal values to the different modeling approaches for projecting future emissions (described in Sect. 4.4.1) based on the quantity of Earth observation data they can incorporate to project future emissions. The historical average modeling approach was assigned a value of "1" because it can be calculated using only 2 years of Earth observation derived data: a map from the starting year and another from the ending year of the historical assessment period. The temporal regression modeling approach was assigned a value of "2" because increasing the *timeframe* and *#maps* can increase the model's prediction accuracy (e.g., R^2 value). Finally, the spatiotemporal modeling approach was assigned a value of "3" because increasing the *timeframe* and *#maps* and incorporating ancillary spatial data sets (which may also be derived from Earth observation data) can increase the model's prediction accuracy. Other approaches for projecting future emissions have not been used thus far for FREL/baseline generation, although some FRELs/baselines were slightly modified to account for anticipated changes in deforestation drivers (e.g., by including a multiplier to increase the projected future emissions).

14.2.3   Comparison of Metric Values
at National/Subnational and Project Levels

As discussed in Sect. 4.4.1, there are typically more specific guidelines that must be followed at the project level compared to at the national/subnational level. To assess whether this resulted in significant differences in the values of any metrics, we compared the metric values at the two different levels using the nonparametric Wilcoxon rank sum test ($p < 0.05$) (Wilcoxon 1945). This test was selected because the metric values were not parametric (values at the project level are determined at least in part by the methodological guidelines that vary depending on a project's location). Chile was excluded from this comparison because its data quantity was not uniform (varied by region).

14.2.4   Relationship Between Earth Observation Data
Quality/Quantity and Methods Used
for FREL/Baseline Generation

To evaluate the relationship between Earth observation data quantity (*timeframe* and *#maps*) and the *#activities* and *model_complexity*, we performed ordered logistic regression modeling at both the national/subnational level and the project level. The Earth observation data quality (*spatial_res*) was excluded from this regression modeling because it was uniform for all countries and projects, as discussed later in Sect. 4.4.3.1. The full regression model for each dependent variable is:

$$\#activities \sim timeframe + \#maps \tag{14.7}$$

and,

$$model\_complexity \sim timeframe + \#maps \tag{14.8}$$

In regards to *model_complexity*, at the national/subnational level a simple logistic regression model was used because the dependent variable had only two values (1 or 2).

To explain *#activities* and *model_complexity* at the national/subnational and project levels, respectively, we selected the best among four different regression models: a null model (no explanatory variables), a model using only *timeframe*, a model using only *#maps*, and a full model using both *timeframe* and *#maps*. Model selection was done based on the Akaike Information Criterion (AIC) (Akaike 1974; Crawley 2005). AIC is a measure of the relative goodness of fit of a statistical model and allows for comparisons among nested models. It is calculated considering the tradeoff between the bias and variance in model construction, and given by:

$$\text{AIC} = 2k - 2\ln(L) \tag{14.9}$$

where k is number of model parameters and ln (L) is the log-likelihood for the statistical model. The model with the lowest AIC value has the highest quality among the models compared. For more information on AIC, readers are referred to a recent review article by Aho et al. (2014).

As the number of samples was relatively low at both analysis levels in our study, we tested the significance of the potential logistic regression models using the parametric bootstrap approach. Bootstrapping entails randomly drawing (with replacement) from the samples to construct a "bootstrapped" set with the same size as the original sample size (some of the original samples may be drawn more than once or not at all in the bootstrapped set), and it is often used to calculate the bias and standard error of a model (Efron and Gong 1983). In this study, the *#activities* or *model_complexity* of each nation/subnation or project were drawn from for the bootstrapping and we calculated the differences between the log-likelihood of the bootstrapped logistic regression model and that of the null model. We performed the bootstrapping 1000 times and compared the log-likelihood difference for the bootstrapped data set with that of the original data set in each iteration. The logistic regression model results were considered to be significant if the log-likelihood difference for the bootstrapped data was smaller than that of original data set in more than 95% of the iterations. All of the statistical analyzes in this section and in Sect. 4.4.2.3 were conducted using R 3.2.5. Chile was excluded from the analysis in this section due to its nonuniform data quantity.

14.3   Results

14.3.1   Quality/Quantity of Earth Observation Data Used

The values of Eqs. 14.1–14.6 at the national/subnational level are shown in Table 14.1, while the values and applied methodologies at the project level are shown in Table 14.2.

At the national/subnational level:

- *%mapped* ranged from 22%–100% with a median value of 100%;
- *timeframe* ranged from 8–22 years with a median value of 13 years;
- *#maps* ranged from 2–15 with a median of 3.5;
- *spatial_res* was 30 m for all countries (all used Landsat data, although some countries also used higher resolution imagery);

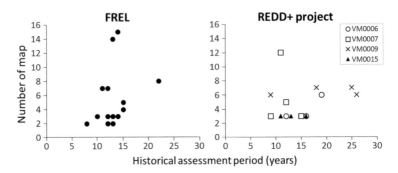

Fig. 14.1 Relationship between historical assessment period and number of maps used to project future emissions in FRELs and VCS-certified REDD+ projects

- *#activities* ranged from 1–5 with a median of 1 (typically deforestation);
- *model_complexity* ranged from 1–2 with a median of 1.

At the project level, one of the four methodologies listed below (Verified Carbon Standard 2016b) were used in all of the 15 projects:

VM0006	Methodology for Carbon Accounting for Mosaic and Landscape-scale REDD Projects
VM0007 REDD+	Methodology Framework (REDD-MF)
VM0009	Methodology for Avoided Ecosystem Conversion
VM0015	Methodology for Avoided Unplanned Deforestation

As can be seen in Fig. 14.1, there were some clear differences in the Earth observation data quality/quantity between projects using different methodologies. As one example, the VM0015 projects all used three maps, while the VM0009 projects used between six and eight maps. However, for the projects in general:

- *%mapped* was always 100%;
- *timeframe* ranged from 9–26 years with a median of 13 years;
- *#maps* ranged from 2–15 with a median of 3;
- *spatial_res* was always 30 m;
- *#activities* ranged from 1–2 with a median of 1; and
- *model_complexity* ranged from 1–3 with a median value of 2.

Regarding the test of correlation between *timeframe* and *#maps,* we found that the two metrics were not significantly correlated at either the national/subnational or project levels ($p < 0.05$, Spearman's rank correlation test; scatterplot shown in Fig. 14.1).

Table 14.3 Akaike Information Criterion (AIC) values of regression models explaining the *#activities* and *model_complexity* at the national/subnational and project levels. Lower AIC values indicate a higher quality model

Explanatory variables	#activities		model_complexity	
	National/subnational	Project	National/subnational	Project
Full model (*timeframe* and *#maps*)	29.138	20.172		45.855
timeframe	29.426	18.639		44.225
#maps	29.503	18.653		43.986
Null model	**28.528**	**17.012**		**42.269**

14.3.2   Comparison of Metric Values at National/Subnational and Project Levels

Comparing the metric values of national/subnational and project levels, we can see that quite similar data and methods were used at both levels. There were no differences in Earth observation data quality (all used 30 m resolution data). In terms of Earth observation data quantity, there were no significant differences in the *timeframe* or *#maps* ($p < 0.05$, Wilcoxon rank sum test). In terms of how the data were used, there was no significant difference in *#activities* ($p < 0.05$, Wilcoxon rank sum test), while *model_complexity* at the project level was significantly higher than that at national/subnational level ($p < 0.05$, Wilcoxon rank sum test).

14.3.3   Relationship Between Earth Observation Data Quality/Quantity and Methods Used for FREL/Baseline Generation

Results of the regression analyses indicated that *#activities* and *model_complexity* were not well explained by the Earth observation data quantity (Table 14.3), as the AIC values were lowest for the null models in all cases. Due to the high uniformity of *model_complexity* at the national/subnational level (all countries except Peru had the same value, as shown in Table 14.1), regression analysis could not be done at this level for this variable. For the other cases, the full model was worst at explaining both *#activities* and *model_complexity* at the project level, but was second best at explaining *#activities* at the national/subnational level. The parametric bootstrap tests also showed that the models that included the Earth observation data quantity variables (full models, and models with either *timeframe* or *#maps*) did not explain significantly better than the null models ($p < 0.05$).

14.4 Discussion and Conclusions

The results provide an overview of the Earth observation data quality/quantity being used for REDD+ at the national/subnational and project levels. Most countries and projects have covered 100% of the county/project area using Earth observation data, used a historical assessment period of at least 13 years, produced at least three maps over the assessment period, and used 30 m spatial resolution data. Synthetic Aperture Radar (SAR) data are often suggested as a complementary data source to optical data for forest monitoring and REDD+ (Reiche et al. 2016) (but have not been used thus far for REDD+), and free approx. 30 m resolution SAR data sets have become available in recent years (e.g., 25 m resolution ALOS PALSAR-1/-2 mosaic data from 2007), so if the median values calculated are considered as minimum requirements then PALSAR-1/-2 mosaic data could be used as a complementary data source starting from the year 2021. The Committee on Earth Observation Satellites (CEOS) provides a searchable online database with details of the past, current, and future Earth observation satellite missions of its member agencies, which could further aid these types of searches (Committee on Earth Observation Satellites 2016). It should be noted that aside from the metrics calculated, other metrics could be computed to obtain even more information on Earth observation data quality/quantity.

Comparing the data/methods used at the national/subnational and project levels, it was found that quality and quantity of the Earth observation data used at the two levels did not differ significantly despite the fact that projects covered smaller areas (and thus are likely to be easier to acquire and process Earth observation data for). The methodological guidelines at the project level may have somewhat constrained these differences (e.g., methodology VM0015 states that the historical assessment period should be no more than 10–15 years, so even if Earth observation data for a longer period are available and easy to acquire, they cannot be used). The number of REDD+ activities assessed using Earth observation data at the two levels was also not significantly different. However, the level of model complexity was significantly higher at the project level than at the national/subnational level, and this was likely due to the stricter methodological requirements at the project level and possibly the greater ease of acquiring/managing/processing ancillary spatial and nonspatial data at the project level (due to the smaller scale).

In terms of the relationship between Earth observation data quality/quantity and the methods used to generate FRELs/baselines, we found that neither the data quality (all countries/projects used 30 m data) nor the data quantity were strong determinants of the number of REDD+ activities considered for FRELs/baselines nor the level of complexity of the models used to project future emissions. Although a larger sample size would of course give higher confidence to the analysis (there is a chance that the null model was selected by AIC because of the limited sample size), the results of this study generally suggest that countries/projects are selecting which REDD+ activities to consider and which modeling approaches to use based on factors unrelated to Earth observation data quality/quantity. In regards to the number of activities considered, it is likely that countries/projects are focusing their efforts for now on the REDD+ activity (or the few activities) responsible for the highest level of greenhouse gas emissions, which is typically deforestation, even if enough data are available for assessing other activities.

In regard to the modeling approach selected for projecting future emissions, while there are strict requirements on the models that can be used at the project level, there is great flexibility at the national/subnational level, so it is possible that countries have other motivations besides model prediction accuracy for their selection of a particular modeling approach. For example, a country may opt for a model that predicts higher levels of future emissions to make it less likely that their actual future emissions exceed these predicted levels. For countries with decreasing rates of emissions in more recent years (e.g., Brazil, Indonesia, and many others), the simplest modeling approach—use of the historical average emission rate—will predict higher future emissions than the more complex modeling approaches that take into account the downward trend in emissions, so these countries may be reluctant to use the more complex modeling approaches. In contrast, for countries with increasing rates of emissions in more recent years (e.g., Peru), the more complex modeling approaches that take into account the upward trend in emissions will project higher future emissions, so these countries may be more willing to select a more complex modeling approach. If the goal is that, as Earth observation data quality/quantity increase over time, the methods used to project future emissions for REDD+ also improve, there may be a need for more specific guidance to be given on how countries facing different national circumstances should project their future emissions.

References

Aho K et al (2014) Model selection for ecologists: the worldviews of AIC and BIC. Ecology 95:631–636

Akaike H (1974) A new look at the statistical model identification. IEEE Trans Automat Contr 19:716–723

Busch J et al (2009) Comparing climate and cost impacts of reference levels for reducing emissions from deforestation. Environ Res Lett 4:044006

Crawley MJ (2005) Statistics: an introduction using R. Wiley, West Sussex, England

Committee on Earth Observation Satellites (2016) The CEOS Database: Mission, Instruments and Measurements. http://database.eohandbook.com/index.aspx. Accessed 30 Mar 2016

Efron B, Gong G (1983) A leisurely look at the bootstrap, the jackknife, and cross-validation. Am Stat 37:36–48

Hamrick K, Goldstein A (2015) Ahead of the curve: state of the voluntary carbon markets 2015. http://forest-trends.org/releases/uploads/SOVCM2015_FullReport.pdf. Accessed 7 Dec 2016

Hilker T et al (2009) A new data fusion model for high spatial- and temporal-resolution mapping of forest disturbance based on Landsat and MODIS. Remote Sens Environ 113:1613–1627

Huettner M et al (2009) A comparison of baseline methodologies for "Reducing Emissions from Deforestation and Degradation". Carbon Balance Manag 4:4

Intergovernmental Panel on Climate Change (2006) Agriculture, forestry, and other land use. In: 2006 IPCC guidelines for national greenhouse gas inventories. http://www.ipcc-nggip.iges.or.jp/public/2006gl/. Accessed 7 Dec 2016

Johnson BA et al (2016) Characteristics of the remote sensing data used in the proposed UNFCCC forest reference emission levels (FRELs). Int Arch Photogramm Remote Sens Spat Inf Sci XLI-B8:669–672

McDermott CL et al. (2012) Operationalizing social safeguards in REDD+: actors, interests and ideas. Environ Sci Policy 21:63–72

Reiche J et al (2016) Combining satellite data for better tropical forest monitoring. Nat Clim Change 6:120–122

UNFCCC (2007) Decision 2/CP.13

UNFCCC (2009) Decision 4/CP.15

UNFCCC (2010) Decision 1/CP.16

UNFCCC (2011) Decision 12/CP.17

UNFCCC (2013a) Decision 11/CP.19

UNFCCC (2013b) Decision 13/CP.19

Verified Carbon Standard (2013) VCS module VMD0007 REDD methodological module: estimation of baseline carbon stock changes and greenhouse gas emissions from unplanned deforestation (BL-UP) version 3.1. http://database.v-c-s.org/sites/vcs.benfredaconsulting.com/files/VMD0007%20BL-UP%20v3.2.pdf. Accessed 8 Dec 2016

Verified Carbon Standard (2016a) VCS project database. http://www.vcsprojectdatabase.org/#/home. Accessed 2 Mar 2016

Verified Carbon Standard (2016b) Find a methodology. http://www.v-c-s.org/methodologies/find. Accessed 15 Apr 2016

Wayne DW (1990) Spearman rank correlation coefficient. Applied nonparametric statistics. PWS-Kent, Boston, USA, pp 358–365

Wilcoxon F (1945) Individual comparisons by ranking methods. Biometrics Bull. 1:80–83

Author Biographies

Brian Alan Johnson received his Ph.D. in geosciences from Florida Atlantic University in 2012 and his M.A. in geography from the same university in 2007. He is now a researcher at the Institute for Global Environmental Strategies (IGES) in Hayama, Japan. His research interests in remote sensing are related to land use/land cover mapping, change detection, and data fusion.

Henry Scheyvens completed his doctorate in political science at Monash University and has served at the Institute for Global Environmental Strategies (IGES) since 2004. He is the area leader for natural resources and ecosystem services research at IGES, where he is leading research on initiatives to mitigate climate change through sustainable forest management.

Hiromitsu Samejima completed his doctorate in ecology at Kyoto University and has served at the Institute for Global Environmental Strategies (IGES) since 2015. Previously, he worked at the Center for Ecological Studies, the Graduate School of Agriculture, and the Center for Southeast Asian Studies, all at Kyoto University. At present he is engaged in research on reducing emissions from deforestation and forest degradation in developing countries (REDD+), sustainable forest management, legal timber trading, and ecology, mostly in Southeast Asian countries.

Chapter 15
Evaluation of Space Programs: Select Findings from the OECD Space Forum

Claire Jolly

15.1 Introduction

Natural resource sustainability, safety, and transport efficiency are quickly growing concerns. Over the years, a number of space applications have demonstrated their usefulness as technical and scientific tools. Despite this, it cannot be taken for granted that public and private investments in space systems will automatically be forthcoming, especially in the current economic context. What would be required to ensure that adequate levels of investment are ultimately secured? The answer, first and foremost, is a sound set of tools to help policy makers arrive at investment decisions. This paper provides a brief introduction to benefit and cost methods used in the space sector, as analyzed in the context of the OECD Space Forum.[1]

In cooperation with the space community, the OECD Space Forum was established to help governments, space-related agencies, and the private sector better identify the statistical contours of the space sector, while investigating the space infrastructure's economic significance, its role in innovation, and potential impacts for the larger economy. This unique international platform contributes to constructive dialogue between stakeholders and the exchange of best practices. The Forum's Steering Group includes the major space-related organizations from OECD economies, from Canada, France, Germany, Italy, Korea, Mexico, Norway, Switzerland, the U.K., and the USA, as well as the European Space Agency (ESA).

---

[1]To learn more about OECD Space Forum activities, see: https://www.innovationpolicyplatform. org/oecd-space-forum/.

C. Jolly (✉)
Organisation for Economic Co-operation and Development, 2, rue André Pascal,
75775 Paris Cedex 16, France
e-mail: claire.jolly@oecd.org

M. Onoda and O.R. Young (eds.), *Satellite Earth Observations and Their Impact on Society and Policy*, DOI 10.1007/978-981-10-3713-9_15

15.2   What Do We Want to Evaluate?

Space systems provide an interesting paradox: they are often considered and funded as research and development programs, but act in many cases as key infrastructures delivering unique public and private services, particularly Earth observation systems. When measuring socioeconomic impacts, basic definitions of what to measure vary.

15.2.1   Space Programs

The first full-scale space programs date from the late 1940s to early 1950s. From the start, they consisted of R&D projects to develop technologies and know-how to send objects into space and utilize this new dimension for science and security purposes. Today, institutional space programs worldwide still cover a wide range of technologies (i.e., launchers, satellites, space stations, ground segment) and disciplines (e.g., telecommunications, Earth observation, navigation, and astronomy), sometimes with "accompanying" programs to involve new users (e.g., commercialization of technologies outside the space sector). Space programs are usually undertaken nationally via dedicated agencies, but also often within a bilateral or multilateral international cooperation framework, particularly in the European context. Since 1980s, a number of private actors have conducted their own space programs directly, for profit (e.g., telecommunications satellite operators, commercial launch providers), but always within a regulatory framework put in place by governments (OECD 2005).

15.2.2   Space Applications

"Applications" are the resulting outcomes of many space programs. Sometimes they are actively sought, to develop specific space products and services (e.g., satellite television); on occasion such results are accidental. The data derived and/or signal issued from a large number of programs initiated for purely scientific purposes can be deemed relevant by large communities of users. Today the value chains for space applications vary, depending on the commercial or scientific benefits of the data or signal provided. That is where the distinction between pure R&D programs (set up for a limited period) and applications (often to be set up on an enduring operational basis) becomes blurry at times. It has been historically difficult to shift programs from the science and technology environment to the financially sustainable operational environment.

15.2.3 Space Infrastructure

The term "space infrastructure" encompasses all systems, whether public or private, that can be used to deliver space-based services. These include both the space and ground segments. As identified in OECD (2005), there are two complementary and interlinked space-based infrastructures. The first one focuses on the "front office", i.e., the one that is "user-oriented" and designed to provide information-related services including communications, navigation signals, and Earth observation data to governments and society at large. The second concerns the essential enabling "back office", i.e., the space transport, satellite manufacturing and servicing infrastructure.

15.3 Tracing Benefits to Satellites

There are different ways to try and trace socioeconomic benefits from satellites' unique capabilities. One can focus on those capabilities (links, signals or data) or base the analysis at the programmatic or infrastructure level. There are already many efforts underway to try and assess the economic impacts of science and R&D programs. As identified by the OECD Space Forum, some of the methodologies used are still evolving. Applying existing techniques to space programs provides some interesting lessons learned. Two examples are provided below.

The macroeconomic approach is often used in the case of large R&D programs or infrastructure to provide cost–benefit information, via economic input–output analyses. The main objective is to measure the growth of productivity in a region or country generated by the investment. Input–output analysis specifically shows how industries are linked together through supplying inputs for the output of an economy. Factors that can be used to construct indicators of productivity include employment, expenditures, income, production of goods and services, and competitiveness. Such factors are of interest at both the national and regional levels. Results of these analyses are derived from macroeconomic data such as changes in GDP, which can then be compared to changes in capital. The challenge when interpreting the material is to find the causal linkages between the program/infrastructure investments and the rise in productivity.

However, the findings of these studies are sometimes contentious. They are also highly dependent on the choice and evaluation of appropriate variables over long periods, as well as the calculations used to assess their cause and effect mechanisms. As an example, the U.S. Federal Aviation Administration's (FAA) Office of Commercial Space Transportation published a report in 2006 on the impacts of commercial space transportation and related industries in other economic sectors, specifically in terms of revenues and jobs that are generated. The economic impact analysis used an input/output method and the Regional Input-Output Modeling System (RIMS II) developed by the U.S. Department of Commerce, Bureau of Economic Analysis. The space sector, as defined in the study by the FAA, was found to be responsible—via direct, indirect, and induced impacts—for USD 98 billion in economic activity in 2004 and 551,350 derived jobs throughout the U.S. All major U.S. industry sectors were affected positively to some extent (e.g., the information services sector, manufacturing, finance and insurance, healthcare and social assistance). As a comparison, using the same methodology the economic impact of the civil aviation industry was found to be over ten times that of commercial space transportation and enabled industries. Methodology-wise, input–output analyses are valuable methods to measure economic impacts. On the other hand, one inevitable drawback of this type of analysis stems from the lack of precise space sector statistics, since the statistical codes used for the study by definition cover more than just space activities (OECD 2016).

At the other end of the analytical spectrum, microeconomic analysis studies the behavior of individual organizations, firms, and customers and their interactions, usually determined by market demand and supply. The use of supply and demand curves is, however, not always directly applicable to space systems and their derived applications because of immature products (new technologies) and non-quantifiable demand. A real technical limitation of microeconomic analysis is the daunting task of assessing accurately all the markets liable to be affected by a specific space technology, and not just when it is innovative. Numerous studies of "spin-offs" have been conducted in the U.S. since the 1960s (such as outputs from NASA's Apollo program), notably of the transfers from space-related hardware and know-how to other sectors (e.g., medical imagery). The value of spin-offs is, however, not easily quantifiable, although they provide interesting illustrations of the diffusion of space technologies in different economic sectors.

A combination of macro and micro approaches tends to provide better estimates, although it will still fail to address potentially larger noneconomic impacts. The choice of a specific analytical technique for impact assessment is not random but context-specific; this is particularly true when looking at space applications. The timing and objective of the assessment, as well as the nature and scope of the public R&D funded, are factors that must also be borne in mind when selecting an analytical technique from an existing toolbox.

15.4 The Way Forward

One important lesson learned from existing evaluations of space programs is that more work is needed to provide evidence-based information to decision-makers and citizens on the benefits (and limitations) of space applications. Existing methods provide useful hints at actual socioeconomic benefits derived from the space infrastructure, particularly for Earth observation, but with diverse inherent caveats, so there is a need to refine further quantifiable analytical tools. In that context, it remains key to maintain the effort in building the international knowledge base on impact assessments to provide know-how and valid experiences to practitioners (avoiding reinventing the wheel). The OECD Space Forum contributes to this activity and will continue engaging actors in the space community and beyond to explore the broad economic and social dimensions of space-based applications.

References

OECD (2005) Space 2030: Tackling society's challenges. http://www.oecd.org/futures/space/space2030tacklingsocietyschallenges.htm. Accessed 8 Dec 2016

OECD (2016) Handbook on measuring the space economy. http://www.oecd.org/futures/oecdhandbookonmeasuringthespaceeconomy.htm. Accessed 8 Dec 2016

Author Biography

Claire Jolly is the head of the Organisation for Economic Co-operation and Development (OECD) Space Forum in the OECD Directorate for Science, Technology and Innovation. Claire has 18 years of experience in business and policy analysis, having worked for both public and private organizations in aerospace and defense, in Europe and North America, before joining the OECD in 2003. Her dual background is in international relations and finance (M.A. Versailles and Cornell universities) and aerospace engineering (ENSTA ParisTech), with a focus on space applications (M.Sc., International Space University, Strasbourg). She is an alumna of the Institute for Advanced National Defence Studies in Paris (Institut des Hautes Études de Défense Nationale, IHEDN). In cooperation with the space community, the OECD Space Forum aims to assist governments, space-related agencies, and the private sector to better identify the statistical contours of the space sector, while investigating the space infrastructure's economic significance, innovation role, and potential impacts for the larger economy. As of September 2016, the Space Forum's Steering Group includes ten members.

Part V
Prospects and Conclusions

Chapter 16
Integrating Earth Observation Systems and International Environmental Regimes

Olav Schram Stokke and Oran R. Young

16.1 Introduction

How can we integrate Earth observation systems and international environmental regimes to enhance the role of satellite observations in solving a variety of large-scale environmental problems?[1] As the capacity of satellite Earth observation to address environmental issues has grown, needs have arisen for enhanced coordination among major players active in this field. We noted in Chap. 1 that particularly important developments in this regard include the growing capacity of satellite observations to (i) provide crucial information on progress in meeting the objectives of issue-specific governance systems (e.g., rates of sea level rise or ocean acidification in the case of the climate change regime or rates of land degradation in the case of the regime to combat desertification); (ii) assist efforts to prevent or mitigate environmental disasters, through measures such as ice warning systems for maritime shipping, flash-flood alerts, or growth mapping of major crops such as rice, maize and soybeans; and (iii) verify compliance with regulatory arrangements

---

[1]International regimes are institutional arrangements created by states to address issues such as international trade, global warming, or fisheries management; see Chap. 1.

O.S. Stokke (✉)
Department of Political Science, University of Oslo, Blindern, PB 1097
0317 Oslo, Norway
e-mail: o.s.stokke@stv.uio.no

O.S. Stokke
The Fridtjof Nansen Institute, Lysaker, Norway

O.R. Young
Bren School of Environmental Science and Management,
University of California, Santa Barbara, CA 93106-5131, USA
e-mail: oran.young@gmail.com

© The Author(s) 2017
M. Onoda and O.R. Young (eds.), *Satellite Earth Observations and Their
Impact on Society and Policy*, DOI 10.1007/978-981-10-3713-9_16

on the part of individual regime members or actors subject to the rules of various regimes.

This chapter explores in detail the question of whether these capabilities have progressed to the point where there is a compelling case to create an integrated environmental Earth observation regime complex and for establishing rules and procedures to guide interactions between providers of satellite observations and users responsible for implementing a range of international environmental agreements. To this end, we analyze organizational matters to be considered for strengthening connections between existing institutions and explore a number of normative issues associated with the development of such arrangements. We conclude with a discussion of next steps regarding the establishment of an international environmental Earth observation regime or a more coherent regime complex.

16.2 Why Focus on Institutional Complexes?

Alongside the proliferation of institutional arrangements in global governance and their rising density and scope, scholars have deepened our understanding of how separate institutional arrangements interact, overlap, complement, or interfere with each other in a variety of ways (e.g., Young 1996; Young 2002; Oberthür and Gehring 2006; Oberthür and Stokke 2011; Orsini et al. 2013). Frequently, governance of a particular issue area is best understood as the result of the interplay of several relevant institutions, with individual institutions affecting the contents, operations, or consequences of other institutions, whether at the same or at different levels of governance. Here, we explain why interplay within larger complexes of institutions is important in thinking about the role of Earth observation systems in addressing large-scale environmental problems.

Studying institutional complexes entails taking an aggregate view of institutional interplay. Terminology differs but considerable agreement exists regarding the conceptual core and empirical significance of this phenomenon. Among the terms in use are "clusters" (Young 1996; Oberthür and Gehring 2006), "governance architectures" (Biermann et al. 2009), "regime complexes" (Raustila and Victor 2004), and "institutional complexes" (Oberthür and Stokke 2011). The conceptual core is clear. Interplay involves interactions among institutions that are distinct in terms of membership and decision making, yet deal with the same activity, or aspects of the same activity, usually in a non-hierarchical manner. This formulation is compatible with Raustiala and Victor's often cited definition of a regime complex as a set of "partially overlapping and non-hierarchical institutions governing a particular issue-area" (Raustila and Victor 2004), but it does not preclude normative hierarchy. Although international organizations and treaties are not usually vertically ordered, elements of formal or actual subordination nevertheless may exist (Stokke and Oberthür 2011).

Among the advantages of taking an aggregate view of institutional interplay is that it directs attention to the distinctive capacities of each institution involved and to the ways in which several institutions may complement one another in the overall governance of an issue area. In international environmental governance, the institution best placed to generate scientific knowledge about the effects of various management programs is often different from the institution with competence to establish such programs or enforce their regulations. To illustrate, the distinctive qualities that provide the International Council for the Exploration of the Sea (ICES) with a central role in the provision of advice on North-East Atlantic fisheries are its membership, comprising the national marine science organizations of most littoral states, and its procedures developed over a period of nearly a century for insulating the advisory function from political pressure without ceding relevance to decision making (Schwach 2000). In contrast, actual regulation typically rests with narrower international bodies whose scope of membership is defined by the relationship of the stocks in question to the economic zones or high seas areas in the region (Stokke 2015). Effective compliance control may require yet another institution, because the membership of all major port states in the region covered by the North-East Atlantic Fisheries Commission (NEAFC) equips this body particularly well for orchestrating the cross-checking of fisher reports with delivery reports, port inspections, and data derived from its satellite-based Vessel Monitoring System, allowing for the integration of real-time tracking of all vessels flagged by member states (Stokke 2014). Each of the institutions involved in this institutional complex retains its operational autonomy, yet their interaction is managed in ways that enhance their combined contributions to overall fisheries governance.

Such retention of operational autonomy without foregoing the realization of gains from coordination makes the institutional complex approach especially relevant for cost-intensive and strategically sensitive activities such as satellite Earth observation. In the remainder of this chapter, we unpack this observation and explore its implications for the development of Earth observation systems to contribute to solving environmental problems.

16.3 The Earth Observation Environmental Regime Complex

Satellites are placed in orbit for a variety of purposes other than Earth observation, such as broadcasting or navigational guidance. Among those satellites dedicated to remote sensing, a majority are equipped primarily to tackle concerns other than environmental problem solving, including national defense and weather forecasting (Adriaensen et al. 2015). Thus, environmental agencies that aspire to tailor the planning and operation of space missions and associated data management to their specific needs must compete with numerous other stakeholders, often with closer and more longstanding relationships to those producing and distributing Earth observation data. This section outlines the Earth observation activities that are relevant to environmental problem solving and considers the complex of institutions that govern them.

16.3.1 Scope of Activities in Focus

It is helpful to divide the chain of activities that connect mission planning and environmental problem solving into three segments: acquisition of data, management of data, and use of Earth observation data for information, assistance, and compliance purposes by those responsible for implementing environmental regimes.

The acquisition of satellite Earth observation data involves a range of closely linked operations including requirement definition, sensor specification and instrumentation packaging, spacecraft launch and flight operations, and data reception (Bailey et al. 2001). Due to the complexity and costs involved, a few leading space agencies, those of the USA, (then) USSR and Japan as well as the European Space Agency (ESA), traditionally handled these operations. Since the late 1980s, however, numerous other nations, especially China and India, have developed state-of-the-art satellite imaging capacity (Morel 2013). According to a recent count, as many as 34 governments have participated in the planning, launching, and operation of some 200 civilian satellites with global land cover observing capacity, roughly half of which were still operational in Belward and Skøien (2015). Commercial Earth observation programs, such as the SPOT satellite system initiated by the French space agency Centre National d'Études Spatiales (CNES) in the 1970s, now supply very high resolution imagery to military as well civilian customers (Morel 2013).[2] In addition, private firms have entered this area of activity, as imaging devices with spatial resolution as fine as 3–5 m have been placed on small and relatively low-cost satellites operated by major companies such as Google or Digital Globe or smaller ones like Planet Labs or Skybox (Belward and Skøien 2015).[3] As a result, a large number of governmental and private actors now engage in the acquisition of Earth observation data potentially relevant to environmental problem solving.

Management of satellite Earth observation data is about distribution and archiving of the received data, often involving further processing to enhance its

---

[2]Here we apply the terminology proposed by Belward and Skøien (2015): very high resolution is less than 5 m; high resolution is 5–9 m; medium resolution is 10–39 m; moderate resolution is 40–249 m; while low resolution means 250 m–1.5 km.

[3]Google acquired Skybox Imaging (from 2016 named Terra Bella) in 2014 as part of the company's efforts to provide real-time images; see OECD (2014).

usefulness for particular purposes (Bailey et al. 2001). The Earth observation data providers certainly engage in these activities. But so do national and international governmental organizations closer to the end users, such as the Global Data Processing and Forecasting System (GDPFS) of the World Meteorological Organization's (WMO) World Weather Watch (WWW) programme (Dillow 2012).

Use of satellite Earth observation data for environmental problem solving, as Part I–IV of this book elaborate, involves a wide range of states and organizations for whom remote-sensing data complement other sources of information that prove helpful when conducting their respective tasks. To illustrate such complementarity, consider how satellite Earth observation data enhance the value of other pieces of information in furthering compliance with international vessel-source pollution rules in the Baltic Sea. Such compliance efforts benefit from the membership of all coastal states in the Paris MOU involving regional port authorities, and in the regional-seas regime for the area centered on the Helsinki Commission (Helcom). Member states conduct some 4000 surveillance flights every year, sometimes integrated through a Coordinated Extended Pollution Control Operation (Tahvonen 2010). In 2011, this coordinated operation involved aircraft and patrol vessels from five regional states as well as satellite surveillance provided by the CleanSeaNet Service under the EU-based European Maritime Safety Agency (EMSA). Since vessels over 300 gross tons engaged in international voyages are obliged under International Maritime Organization (IMO) rules to be equipped with Automatic Identification System (AIS) transponders, backtracking of vessels located near an oil spill can, in conjunction with a regional oil-drift forecast model endorsed by the scientific branch of the regional regime, help identify one or more likely culprits for subsequent port-state inspection (Tahvonen 2010). Thus, Earth observation data blend in with other types of information, helping to narrow the scope of those suspected of rule violation and allowing more efficient use of the port-state control mechanism (Stokke 2014). In remote-sensing terms, the user interprets the processed data conveyed in the satellite images in light of inputs from other sources, yielding "analyzed information" (Smith and Doldirina 2008) in support of environmental problem solving.

Agencies tasked with implementing the provisions of environmental regimes are newcomers among the users of satellite Earth observation data, compared to users of data located in meteorological services or national security agencies. Among the premises of this book is the proposition that there is considerable room for improvement with respect to environmental agency awareness about recent advances in satellite Earth observation technologies and the opportunities they provide for more effective detection and monitoring of environmental problems or for supporting response action or compliance mechanisms. As the next section argues, the converse is also true; the coordinating institutions set up largely by providers and managers of Earth observation data do not provide sufficient exposure to the practices and needs of environmental governance.

16.3.2   Boundaries of the Earth Observation Environmental Regime Complex

The many institutions that co-govern satellite Earth observation activities derive partly from international space law centered on a small number of United Nations (UN) treaties and regional collaboration agreements adopted in the early decades of space exploration, and partly from a wide array of soft-law instruments aiming to support mutually beneficial coordination. With the rising commercialization of space, the more recent of these coordination mechanisms typically take the form of partnerships, involving national space agencies and international organizations, as well as industry associations. Prominent among those partnerships is the Group on Earth Observations (GEO), set up in 2005 to enhance the compatibility of existing observing and processing systems that would, importantly, continue to operate under their own mandates, hence the term Global Earth Observation System of Systems (GEOSS; see Part III).

The subset of space law that is relevant to Earth observation activities deals primarily with acquisition of data. The 1967 Outer Space Treaty, for example, makes clear, consistent with its broader "open-skies" principle, that satellite Earth observation is a legitimate activity that does not require consent from states or private actors observed; Article IX adds only that states are to inform others about activities that might affect their use of space. The 1972 Liability Convention makes states liable for damage caused by space objects launched from their territory, including those launched by private entities. Recently, the Permanent Court of Arbitration proposed more elaborate procedures for handling liability complaints relating to space activities (Weeden and Chow 2012). Crucial to the implementation of these various rules for coping with unintended impacts and space congestion, increasingly challenging with the deregulation and privatization of space activities during the past two decades, is the work of the International Telecommunications Union (ITU), a specialized UN agency that collaborates with national space agencies to assign satellite orbits and radio frequencies (OECD 2014; Viikari 2007).

Globally applicable provisions exist also for Earth observation data management, especially those set forth in the UN Principles Relating to Remote Sensing of Earth from Outer Space, adopted in 1986. Although that instrument is sometimes referred to as the normative centerpiece for satellite Earth observation activities (e.g., Kuriyama 2005; Smith and Doldirina 2008), it is only a non-binding resolution and its contents are not precise enough to serve as meaningful legal commitments. Nevertheless, the general thrust of the Principles is clearly favorable to those who wish to align Earth observation data provision and management to the needs of environmental problem solving. It starts out by defining remote sensing as "sensing of the Earth's surface from space … for the purpose of improving natural resources management, land use and the protection of the environment" (Principle I), adding that such activities are to be "carried out for the benefit and in the interest of all countries" (Principle II). Hortatory language and qualifications appear as soon as norms become specific to actual management of Earth observation data. States are "encouraged to provide for establishment and operation of data collecting and storage stations and processing and interpretation facilities … wherever feasible" (Principle VI), rendering Earth observation data accessible in a non-discriminatory manner and at reasonable cost (Principle XII). The most prominent soft-law norms on Earth observation data management guide states toward open-access practices but, as Smith and Doldorina (2008) point out, they are silent on the role of private companies that are steadily becoming more important in space affairs.

Actual practices regarding Earth observation data management have varied considerably among states and organizations depending on purpose and fineness of resolution. The meteorological and the oceanographic communities opted for open and free access in the 1990s (Harris and Baumann 2015); this policy was also reflected in the International Charter on Space and Major Disasters, made operational by the major space agencies in 2000 and now providing data and imaging access even to non-parties during a crisis situation.[4] In other domains, data at medium resolution or finer have traditionally come at a cost (Belward and Skøien 2015). Still, the current trend among public data providers is toward free-and-open access. While cost-retrieval policies made US Landsat data subject to payment during the 1980s and 1990s, even by domestic public agencies, a gradual de-commercialization peaked in the 2008 decision to make all Landsat data available to all at no cost (Sawyer and Vries 2012). From 2009, China-Brazil Earth Resources Satellite (CBERS) program data can be downloaded free of charge by all African countries (OECD 2012); more recent policies also include Latin America.[5] A similar movement is evident in the ESA Data Policy, which used to differentiate between research and commercial uses, charging for the latter (Harris and Browning 2003), but no longer does (ESA 2012). Consistent with this trend, the European Commission's Copernicus Data Policy from 2013 provides that non-sensitive data from this program are available on a "free, full and open basis" (EU 2013).[6] Even commercial ventures are adapting their pricing policies: medium resolution SPOT satellite data more than 5 years old are now available free of charge to non-commercial users (Selding 2014).

This move toward more open access to publicly acquired data has been fuelled by a combination of declining prices on satellite imaging, rising concern among legislators that they are "paying twice" for Earth observation data, and growing appreciation that wide availability of such data can empower individuals, civil society, and business and stimulate the economy (Sawyer and Vries 2012). Earth observation data are included among the high-value areas in the Open Data Charter

---

[4]The Charter has been activated more than 500 times since 2000; details at www.disastercharter. org (accessed 30 August 2016).

[5]Statement by China's State Administration of Science, Technology and Industry for National Defence on the occasion of the fourth joint Chinese-Brazilian Earth observation satellite being put into operation, cited in New China, 14 August 2015, http://news.xinhuanet.com/english/2015-07/ 14/c_134412054.htm (visited 4 August 2016).

[6]See especially Article 3 (open dissemination principles) and Articles 11–16 (on conflicting rights and sensitivity).

adopted by the Group of Eight Industrialized Nations (G8), with its openness-by-default principle and its complaint that states "do not always share these data in ways that are easily discoverable, useable, or understandable to the public".[7] On this cue, as the primary international private–public partnership for improving coordination along the Earth observation chain of activities, GEO has agreed to "promote and encourage" data management principles close to the open-access end of the continuum of existing practices. Specifically, GEOSS partners and others should strive to make their Earth observation data and metadata *discoverable* through catalogues and search engines; *accessible* via online services; *usable* through encodings widely accepted among user communities, quality control, and comprehensive documentation of access and use conditions as well as data provenance; *preserved* for future use; and *curated* by means of correction, update, and necessary reprocessing (GEO 2015a).[8] Limitations on the actual usability of available data for environmental problem solving, and how to overcome them, are the subject of the next section.

16.4   Integrating the Earth Observation Environmental Regime Complex

In Chap. 1, we offered a taxonomy of ways in which satellite Earth observations can promote national or international efforts to solve environmental problems. In conjunction with other types of data, satellite Earth observations can help to inform decision-makers about the severity of environmental problems by identifying undesirable conditions, monitoring their development, and assessing the adequacy of efforts to combat the problems. Satellite Earth observations can also be vital for provision of assistance to those exposed to environmental risks or disasters, like search and rescue operations or early warning systems for floods and tsunamis. In addition, satellite observations play crucial roles in certain systems for verifying compliance with international environmental commitments, as illustrated in the Helcom oil pollution case. This section identifies some important initiatives intended to strengthen each of these functions and comments on institutional adaptations that might enhance them.

---

[7]See Preamble and Principle 1 of the G8 Open Data Charter, available at https://www.gov.uk/government/uploads/system/uploads/attachment_data/file/207772/Open_Data_Charter.pdf (accessed 2 August 2016). The G8 refers to a group of highly industrialized countries that meet annually on important economic or other issues, comprising Canada, France, Germany, Italy, Japan, Russia, the UK, and the U.S., plus the EU.

[8]Metadata is information about the data, including its format and provenance, i.e. how it was acquired, the extent of quality control and curation, etc.

16.4.1 Information

As Chap. 1 brings out, cases abound in which satellite Earth observations help to identify and monitor environmental problems or assess progress toward solving them. In some of these cases, institutional mechanisms exist and allow interaction among providers and users but too often such interaction occurs at a stage when decisions regarding instrumentation and data processing capacities have already been made.

Especially in the climate and biodiversity areas, environmental users of satellite Earth observation data have made progress in coordinating their interactions with data providers. An important example features the 50 Essential Climate Variables (ECVs), each with its desired requirements for accuracy and spatial and temporal resolution, defined by the Global Climate Observing System (GCOS) in close interaction with the Intergovernmental Panel on Climate Change (IPCC) and UN Framework Convention on Climate Change (UNFCCC) (Bojinski et al. 2014). These ECVs are also applicable to a range of adjacent environmental issues, such as air quality monitoring and forecasting as well as water resource management (Tan 2014). A corresponding set of Essential Biodiversity Variables (EBVs) are being developed under the GEO Biodiversity Observation Network (BON), in response to the Aichi Targets for 2020 defined under the UN Convention on Biological Diversity (UNCBD) (Pereira et al. 2013). The BON initiative links leading space agencies with scientific and governmental actors at national and international levels, including the Intergovernmental Science-Policy Platform on Biodiversity and Ecosystem Services (IPBES). Among its ongoing projects is to develop a "BON in a Box" regionally adaptable toolkit aiming to harmonize local, national, or regional biodiversity monitoring initiatives.[9] These and other initiatives aim at making satellite Earth observation data relevant to a broader set of environmental monitoring efforts than hitherto.

Among the strengths of the essential variable approach is to provide focal points for Earth observation data providers when planning missions and instrumentation payloads. In many cases, benefitting fully from satellite capabilities will require recurrent interaction among users and those responsible for mission planning and subsequent data processing. Such recurrent interaction will be particularly important for the ongoing efforts to develop essential variables for monitoring and assessing marine ecosystems, in part because biological and ecological characteristics interact in ways that are often complex, poorly understood and, especially relevant to our discussion, varying in importance seasonally and over time (Hayes et al. 2015). To illustrate, high up on the list of candidate essential marine ecosystem variables is satellite-observed, color-variant chlorophyll concentration, indicative of phytoplankton density and primary production more generally. However, the algorithm used to convert ocean color into estimates of standing

---

[9]On the GEO BON initiative, see https://www.earthobservations.org/activity.php?id=35 (accessed 8 August 2016).

stocks of phytoplankton and primary production must take into account time and space variant factors, such as relative abundance of species, the layering of chlorophyll in the water column, and in some interesting locations, the standing stocks of algae in and under the ice (Constable et al. 2016). The intensity and recurrence of provider–user interaction needed to optimize the role of satellite data in an overall monitoring effort, taking into account the relative costs of alternative acquisition methods, is likely to vary considerably among environmental regimes.

While the most well-known examples of satellite observations supporting assessment of progress in environmental governance are found in the ozone and climate areas (see Chap. 1), some innovative providers are exploring new or more extensive applications of existing data and observing capabilities. Consider the example of the Global Mangrove Watch, developed by the Japan Aerospace Exploration Agency (JAXA) in partnership with, among others, the global NGO Wetlands International and offered as a tool for the Global Wetland Observing System under the Ramsar Convention (Rosenqvist et al. 2014). Trends in mangrove cover can be highly relevant also for assessing progress or setbacks under UN conventions on climate change (UNFCCC), biodiversity (UNCBD) and conservation of migratory species of wild animals (Convention on the Conservation of Migratory Species of Wild Animals, CMS) (Lucas et al. 2014). The JAXA initiative exploits the longevity of global-scale remote sensing at frequencies conducive to mangrove observation in combination with rapidly advancing data processing capacity. Such conduciveness for particular environmental applications would be stronger and more effective if a broader range of actual and potential users, spanning the functional ambits of several environmental regimes, were to be consulted at the planning stage of satellite Earth observation missions.

16.4.2 Assistance

Goals pursued under some international environmental agreements fuse with broader societal-security concerns, not least those aiming to prevent and respond effectively to environmental disasters such as floods and wildfires. Satellite-derived information can support each stage of the disaster management cycle, including prevention, warning, response, and recovery. Sweta and Bijker (2013) point out that those engaged in the various stages of the disaster management cycle differ widely as to what they perceive as decisive quality characteristics of Earth observation data and in terms of the extent of metadata needed. This situation suggests that new arenas and more inclusive review procedures would help to improve the fitness of the wide range of available Earth observation products to the distinctive needs of diverse users.

Considering that the meteorological community has been prominent among the non-military users of Earth observation data ever since the dawn of the satellite era (Edwards 2006), it is understandable that flood management is an area with relatively firm institutional links along the provider–user chain, especially with respect to early warning. Geostationary satellites hovering over fixed positions allow meteorological agencies to issue forecasts of where and when thunderstorms are about to hit and where the likelihood of flash floods is particularly high (Tan 2014). WMO has played an active role in integrating satellite and otherwise-obtained data for this purpose, including through its Flood Forecasting Initiative (FFI). This initiative, operational since 2003, builds capacity among meteorological and hydrological agencies in developing countries to compile and disseminate the relevant information in flood-critical situations to those responsible for civil protection and response (Sene 2008).

Important as it is to empower meteorological and hydrological agencies by means of satellite Earth observation data, the overall flood management capacities of exposed regions are likely to benefit even more if patterns of involvement are expanded further. As reflected in a series of soft-law instruments under the 1992 UN Economic Commission for Europe (UNECE) Water Convention (the environmental framework agreement with the longest record of flood-risk responsibility) and in its global counterpart the 1997 UN Watercourses Convention, modern flood protection strategies give pride of place to the water retention capacities of nature itself, including the networks of small waterways, wetlands, soils, and grasslands throughout a river basin (UNECE, 2009).[10] This environmental dimension of flood protection means that flood management must be an integral part of general river basin management plans often developed by international environmental bodies established by riparian states. While WMO is sensitive to this need for integrated flood management, as evident in the Associated Programme on Flood Management (APFM) initiated jointly with the Global Water Partnership (GWP),[11] those responsible for such larger river basin management plans are likely to have Earth observation data needs distinctive from those of meteorologists and hydrologists.

Wildfire prevention and response is another area where satellite Earth observation data are key and where institutionalized provider–user interaction may help to tailor relevant information products to the needs of environmental managers. The environmental and societal challenges posed by wildfires are enormous. In the World Bank's disaster management portfolio, only projects addressing floods and droughts have been more numerous (IEG 2006). One means for linking providers and users in this field is the Fire Program of the Global Observation for Forest Cover and Land Dynamics (GOFC–GOLD) effort, an international partnership including the Committee on Earth Observation Satellites (CEOS; a coordinating body for space agencies) as well as research communities, NGOs, and subsidiary bodies of the UNFCCC and UN Convention to Combat Desertification (UNCCD)

---

[10]See Model Provisions on Transboundary Flood Management, especially Provision 4, reproduced in ECE (2009, Annex I).

[11]More information at the program's website, http://www.apfm.info/ (accessed 16 August 2016).

(Calle and Casanova 2008). Among the projects contributing to this program is the Fire Information for Resource Management System (FIRMS), a set of applications providing access to near real-time web-based fire maps and associated data sets without exceeding the limited download capacities of many developing countries (Davies et al. 2009). While close interaction with focus groups of Earth observation data users in Central America, Africa, and Asia was important to the design of these applications, the project is provider-initiated and interaction with users have revolved around the processing and dissemination of satellite data, not data acquisition.

Efforts to prevent and respond to disasters such as floods and wildfires are increasingly integrated within larger environmental management plans, thus involving national and international environmental agencies with Earth observation data needs that may exceed or differ from those defined by civil protection agencies. Such diversity should be reflected in increasingly institutionalized provider–user interactions that aim to enhance the role of satellite data in supporting disaster management.

16.4.3   Compliance

Involvement of existing and potential environmental management users is no less important when designing new Earth observation products for the third major function of satellite data in environmental problem solving, improving adherence to international commitments. As noted in Chap. 1, many states have used satellite data as part of their national inventory reporting under the Kyoto Protocol, especially with respect to emissions and removals associated with land use and land cover changes (e.g., Reddy et al. 2015). They will continue to do so for pledges under the 2015 Paris Agreement. In other sectors such as fisheries management, satellite observations can also be crucial to verification and review activities, those parts of a regime's compliance system that evaluate the reliability of national or target group reports by providing independent information suitable for cross-checking.

Satellite-based tracking of individual fishing vessels as part of systems for compliance monitoring is a well-established practice in many regional fisheries management regimes. So far, however, these vessel monitoring systems have required specially devised on-board transponders typically linked to flag-state enforcement agencies and have therefore been relevant only for vessels registered with a regime member. The Global Fishing Watch initiative by Google (a relative newcomer among Earth observation data providers) in partnership with certain NGOs interested in Illegal, Unreported, and Unregulated (IUU) fishing, seeks to overcome that limitation by using instead the AIS signals that IMO has mandated for certain vessel types under its Safety of Life at Sea Convention (Ouellette and Getinet 2016; see also Chap. 1).[12]

In order to function as intended, however, the Global Fishing Watch initiative needs complementary action under global and regional institutions responsible for maritime safety or resource management (McCauley 2016), action that is not necessarily forthcoming. IMO mandates AIS transponders only for vessels larger than 300 gross tons and engaged in international voyages, meaning that application to most fishing vessels is discretionary for the flag states. Unfortunately, IUU fishers are frequently registered by flags-of-convenience states unlikely to prioritize compliance. A 2015 submission by two environmental NGOs encouraging a Joint FAO/IMO Ad Hoc Working Group on IUU Fishing to advocate stricter AIS regulations met with a lukewarm response (FAO 2015). Among the counter arguments is the expectation that use of the AIS system for compliance purposes might encourage tampering or even disabling the transponders, undermining the safety-at-sea objectives that motivated the creation of the system. Some states and regional organizations have nevertheless mandated AIS signals for large parts of their fishing fleets. Even if regulatory developments are slow at the global level, this satellite-supported tool may become important in combatting IUU fishing.

As with the information and assistance roles, designing steadily more effective Earth observation products for compliance purposes frequently requires coordination among data providers and users with sufficiently high stakes in the outcome to mobilize political energy for creating an effective institutional environment.

16.5 Organizational Issues

It is neither practical nor efficient for most regimes established to solve specific environmental problems to operate their own Earth observation systems. This is true even for those dealing with global or widespread issues like the depletion of stratospheric ozone or the destruction of tropical forests. What we can expect in this realm is the emergence of a complex of institutions involving providers of Earth observation data that have specialized in the development of capabilities involving the acquisition and management of data together with a range of potential users that require data for different purposes. Those users may want to develop means of integrating Earth observation data with other types of data to maximize their effectiveness in solving the problems that fall within their remit. As a result, it is important to address a series of issues that center on developing productive relations between providers and users of Earth observation data. Partly, this is a matter of achieving a good match between the capabilities of the providers and the needs of the users to maximize success in solving specific problems. In part, however, it is a

---

[12]For details of this Global Fishing Watch initiative, see http://globalfishingwatch.org/ (accessed 30 August 2016).

matter of developing rules governing the provider–user interface to ensure that this relationship conforms to broader societal preferences. Should Earth observation data providers make data available free of charge, for example, or is it legitimate for them to charge for their services? If charges are involved, should users of Earth observation data be allowed to shop among providers in order to obtain the most favorable deals for their purposes?

16.5.1   An Earth Observation Environmental Data Forum

So long as Earth observation data providers and users responsible for implementing environmental regimes operate independently, it is unlikely that the data from providers will dovetail well with user needs. Left to their own devices, providers will design payloads that focus on the latest technologies and cutting-edge capabilities regarding the acquisition of data. For their part, users are likely to formulate their requests for data with little attention to the practical limits of what the providers can supply given realistic cost constraints. What is needed in this regard is a conscious effort to find ways to bring together the providers and the users not only at the stage of distributing available data to individual users but also at the stage of planning regarding the number and types of satellites to launch and the instrumentation packages loaded on individual satellites. In effect, there is a need for an end-to-end relationship in which providers and users engage in cooperative efforts all the way from designing missions to evaluating the results and planning for improved missions at a later stage.

What is the best way to engage in this sort of matchmaking? Assuming that the community of providers and the community of users remain separate groups with their own goals and objectives, there are several strategies that may prove helpful in this regard. One strategy is to match specific providers (e.g., ESA or JAXA) with particular users (e.g., those responsible for administering the ozone regime, the climate regime, or the biodiversity regime). This would allow for intensive engagement focused on the data needs of those seeking to solve specific environmental problems. Another strategy is to focus on the functions we reviewed above (i.e., inform, assist, comply) and to match the capabilities of EO data providers with the sorts of data needed to handle each of these functions.

While these strategies may make sense under some circumstances, they have obvious drawbacks as general approaches to matchmaking between providers and users of Earth observation data. There is considerable overlap among environmental regimes in their needs for data, so it makes sense to look for synergies in responding to these needs and to encourage cross-regime communication designed to make use of Earth observation data in the most efficient way. Similarly, many individual regimes involve needs for data covering the full range of functions we have identified. In such cases, it makes sense to look for ways to coordinate efforts to provide the different types of data needed by regime administrators in an efficient manner.

One interesting response to these concerns would be to develop an informal but nevertheless well-defined and functionally broad forum in which providers and users of Earth observation data relevant to solving environmental problems could interact with each other to discuss issues involving the match or fit between the capabilities of providers and the needs of users. Such a forum would not be a substitute for the development of contractual agreements between specific providers and users covering their terms of engagement (see Sect. 16.5.2). Rather, the Earth observation environmental data forum would provide a more general setting for discussion of emerging capabilities in this field and the evolution of environmental regimes creating new uses for Earth observation data. GEO might play a constructive role in orchestrating the participation of providers in such a forum. An Earth observation environmental data forum would follow up on the call by the GEO Ministerial meeting in 2015 to "strengthen its focus on users and stakeholders and in particular develop new approaches to effectively engage with ... Multilateral Environmental Agreements" (GEO 2015b).

There is no counterpart to GEO on the environmental user side. However, the UN Environment Programme (UNEP), the World Bank, and the Global Environment Facility (GEF) are involved in the implementation and funding of many of the major international environmental regimes; UNEP and the World Bank are also GEO partners. Building on the experiences from endeavors that involve individual international environmental regimes, such as the definition of ECVs, a broader Earth observation environmental data forum could facilitate similar processes among wider sets of users, linking them to relevant provider expertise on advances in technological capabilities and on remaining cost constraints. The scope of participation by those responsible for operating environmental regimes could vary depending on the substantive focus of each forum meeting, which may be defined in terms of issue area (e.g., forestry, fisheries, biodiversity) or function, or any combination of these.

There is no need to formalize relations among these entitics regarding the creation and operation of an Earth observation environmental data forum. But there is an opportunity for the progressive development of social practices in this realm. Deeper involvement on the part of major international funding agencies such as GEF and the World Bank would allow such a forum to benefit from the particular experience and expertise those agencies have in evaluating societal benefits across environmental projects and issue areas. They could also provide financial muscle to fund specific Earth observation activities tailored to their project portfolios—the World Bank is a major user of Earth observation data in support of its investment decisions and project implementation (Royal Society 2015). It might well be appropriate for the GEF to provide the relatively modest funding needed to support periodic meetings of an Earth observation environmental data forum bringing together leading providers and users.

16.5.2   Terms of Engagement

There remains the issue of the contractual arrangements covering relations between providers and users regarding the use of specific data applied to particular environmental problems. Many of those who have thought about the uses of Earth observation data believe that publicly funded agencies (e.g., ESA, JAXA, NASA) should be responsible for the collection of relevant data and that the data they obtain should be made available for environmental problem solving on a free-and-open basis. On the other hand, as we have noted, private companies are entering the ranks of Earth observation data providers; they advertise their services as efficient sources of data needed to handle a variety of functions. These observations raise a number of classic issues regarding the relative merits of public and private provision of various services. Without going into detail regarding these underlying issues, we can identify and comment on several concrete matters relating to the terms of engagement between Earth observation data providers and users in need of particular types of data to address the environmental problems they seek to solve.

Today, many countries (including developing countries like China and India) and a growing number of private corporations operate satellites and use them to collect various types of data on a regular basis. This suggests that users endeavoring to solve specific environmental problems may be tempted to shop around, searching for different types of data to meet their needs from a variety of providers. Nonetheless, both providers and users may find that they have significant incentives to form clear-cut (though not necessarily exclusive) and lasting relationships covering the terms of engagement between them regarding the provision and use of Earth observation data.

For one thing, individual users may wish to have input at the design stage, participating in decisions relating to the instrument packages loaded on specific satellites, the orbits of satellites, or arrangements involving the coordination of two or more satellites to ensure needed coverage, in return for making long-term commitments to using the services of a particular provider. If the problem is simply a matter of monitoring rates of loss of tropical forests or the magnitude of marine dead zones, it may be sufficient to work with low-resolution images that are collected on a weekly or even monthly basis. When the problem is a matter of providing early warning regarding tidal waves likely to hit exposed coasts or forecasting where and when a hurricane will make landfall, there is a need for higher temporal resolution data that can be collected and disseminated rapidly. While providers have coordinating mechanisms of their own (e.g., CEOS and GEO) and users tend to be grouped into distinct regimes dealing with separate issues (e.g., stratospheric ozone depletion, marine pollution, deforestation), there may be good reasons to make a concerted effort to bridge this gap when it comes to addressing specific environmental problems. One way to do this is to move toward the formation of lasting contractual relationships organized around regime complexes encompassing all those players endeavoring to solve more or less well-defined environmental problems.

Users will often want to merge data obtained from a variety of sources, including observations from aircraft, in situ monitoring devices, and mobile sources as well as satellite Earth observation data. In such cases, they will have an incentive to work with multiple providers to develop programmatic arrangements that maximize their ability to carry out monitoring, reporting, and verification functions effectively. Unlike consumers whose engagement with providers is likely to end once they have completed the purchase of a specific product, relationships between providers of data and users dealing with environmental problems will normally be complex and ongoing. This suggests that both sides will find it beneficial to spell out terms of engagement in relatively explicit contractual agreements. There may even be reasons to develop mutually agreed procedures to provide authoritative interpretations of the terms of resultant contracts when the parties disagree about their application to specific situations.

When public agencies are in need of services, ranging from the construction of infrastructure to the provision of security services at public facilities, it is normal to issue requests for proposals and to establish some procedure both to evaluate the merits of the proposals received and to make decisions regarding preferred proposals. The evaluation may include a variety of considerations relating to the track record of the providers and the quality of the services to be provided as well as the proposed budget. For various reasons, this model may not be applicable to the relationships between Earth observation providers and users we are considering in this analysis. Among other things, the providers themselves may be public agencies and there may be questions regarding the extent to which it is appropriate to charge for providing the relevant data. Still, there is merit in thinking in contractual terms about the provision of Earth observation data to address specific environmental problems, especially in cases where the options include turning to private enterprises as sources of supply. Among other things, an approach of this sort can produce well-defined and stable expectations on the part of all participants, a matter of particular importance in relationships that are expected to last indefinitely and to adapt successfully to changing circumstances.

With regard to the specific question of payment for the provision of Earth observation data, there is no simple answer. The current trend in which space agencies increasingly provide Earth observation data on an open-access basis is appealing not only for ideological reasons but also because environmental regimes are seldom well-endowed with funds that would be needed to purchase data. Still, it would be a mistake to jump to the conclusion that free-and-open access is the preferred policy under all conditions. Supporting the growth of national space industries and associated technology development is a major priority for governments active in this sector (OECD 2014). Among ESA member states, for instance, strengthening industrial competitiveness is the most frequently cited motivation for space activities (Adriaensen et al. 2015). It is not surprising, therefore, that policies of open access to data free of charge are questioned and sometimes opposed by companies that specialize in the provision of Earth observation products similar to those increasingly available at no cost. CNES, for example, specifically confined its zero-cost policy to older and medium resolution Earth observation data so as not to undercut markets exploited by its SPOT partner Airbus Defence and Space (Selding 2014). The European Commission has reportedly promised to adjust its Copernicus Data Policy should it jeopardize the viability of European private satellite operators.[13]

Moreover, users often need to order specialized data to address the problems they are seeking to solve. In some cases, private providers may be more responsive to these needs and more efficient in minimizing the costs involved, as illustrated in the role of commercial actors in developing networked systems of smaller and smaller satellites (OECD 2014). More generally, we should not exclude the possibility that private providers have more compelling incentives to develop new or more cost-efficient observing systems and finding ingenious ways to configure them to meet the needs of certain users.

None of this is meant to undermine the premise that open-access data supplied by national space agencies will prove to be the best option under a variety of conditions. But it is clear that we should not simply assume that this option constitutes the preferred strategy in all cases. In efforts to come to terms with the provider–user interface on a case-by-case basis, the issue of financial arrangements must be tackled head on with the objective of arriving at mutually satisfactory arrangements.

16.6   Impacts on Earth Observation Sustainability

Placing environmental institutions more centrally in the larger Earth observation environmental regime complex promises to enhance the role of satellite data in environmental problem solving. But it also involves costs for Earth observation data providers that, for long-term sustainability, must somehow be balanced by substantive provider benefits. Among the costs are not only the time and energy that providers must invest in various processes for identifying the environmental needs that match existing or future acquisition and processing capacities but also the opportunity costs of instrumentation packaging specially tailored for environmental problem solving. To illustrate this observation, we address three impacts of firmer connections among Earth observation data providers and environmental agencies, revolving around economic viability, environmental sustainability, and certain normative or ethical concerns that loom steadily larger as satellite systems become more sophisticated.

---

[13]See Selding (2014): "the Commission may reassess the sensitivity of specific Copernicus data on its own initiative or at the request of a member state"; see also EU (2013), Art. 15.

16.6.1 Economic Viability

The end of the Cold War made it less compelling for governments to allocate resources to space activities for security reasons, as did the rising preparedness of private actors to develop space capacities independent of taxpayer support. These developments have raised questions about the stability of public funding for the major space agencies. In the case of NASA, the political support for its budget has been perceived, in the words of one observer, as "a mile wide but an inch deep" (Handberg 2003; cited in Cobb and WNW 2011). Obtaining the necessary long-term funding for Earth observation activities is seen as particularly challenging since they do not typically provide spectacular "space events", like landings on the moon or explorations of Mars, that have fuelled popular interest in and support for space activities (Morel 2013). Some leading spacefaring nations have seen considerable variation in the public support for spending on space. Surveys indicate that the share of the US public that considers government spending on space activities as excessive rose from about a third in 1988 to around half in the mid-1990s, then leveled off to around 40% (Cobb and WNW 2011).

Despite these worries over the economic viability of space activities, governmental space budgets supporting civilian uses have proven stable among the OECD countries, even during the financial crisis of 2008–2009, whereas emerging economies like China, India, and Brazil have increased their space budgets considerably (OECD 2014). As highlighted by Gaubert (2002), governmental preparedness to pay for space activities derives in part from interest among space-capable states to retain or improve their competitiveness in a technology-intensive and growth-driving sector. A recent review of European national space strategies indicates that only security and transportation loom larger than environment and resource management among priority areas seen as benefitting from engagement in space activities (Adriaensen et al. 2015). New initiatives that broaden and enhance the use of satellite data for environmental problem solving are likely to stabilize and reinforce the perception that public spending on space activities is money well spent, adding to the social capital of those engaged in the space sector.

Indirectly, therefore, enhanced contributions to environmental problem solving are likely to contribute to the financial viability of space-based Earth observation activities, even if the environmental regimes that stand to benefit are not charged directly for these services. Economic benefits for providers can also be more direct, especially when intensive or recurrent provider–user interaction promises to generate specially tailored Earth observation products, enabling more efficient or cost-effective operation of certain governance tasks, such as problem monitoring or compliance control. In such situations, international environmental bodies may be motivated to contribute financially to realize the product, either directly through their own budgets or (perhaps more frequently) through related environmental funding agencies such as the GEF or one of the many funds for climate action. As noted above, active involvement of the GEF and the World Bank in a broader Earth observation environmental data forum would likely promote the ability of international environmental regimes to trigger financial muscle in support of satellite-based Earth observation activities.

16.6.2   Environmental Sustainability

A more prominent role for satellite Earth observation in environmental problem solving would strengthen one strand of a broader effort to improve the environmental sustainability of space activities. The other strand of that effort has revolved around a series of congestion problems in outer space, notably the scarcity of orbits appropriate for geosynchronous or sun synchronous observation, the rising challenge of frequency interference, and the risks space debris poses to safe spacecraft operation, including those resulting from collisions or intentional destruction of satellites (Williamson 2012).[14] The rising prominence of outer space sustainability is evident in the adoption of the 2007 Space Debris Mitigation Guidelines by the UN Committee on the Peaceful Uses of Outer Space (COPUOS) and in that Committee's decision to place outer space sustainability on its agenda from 2010 onwards, having tasked a Working Group on Space Sustainability with developing specific inputs suitable for best-practice guidelines concerning debris, the safety of space operations, and management of the radio-electric spectrum (Brachet 2012).

Like any sector that generates obvious societal benefits but also threatens certain environmental qualities or exploits finite resources, those engaging in private or public space activities must demonstrate awareness of the environmental problems they generate and dynamism in efforts to mitigate them. Developing and exploiting positive externalities by using facilities in space to solve sustainability problems on Earth fits well with the sector's growing preparedness to place negative externalities in outer space such as debris and congestion on the international political agenda.

16.6.3   Normative or Ethical Concerns

Even if the terms of engagement governing interactions between providers and users of Earth observation environmental data are well-defined and widely understood, there are some important normative issues that arise in this realm and that are destined to take on greater prominence as satellites become more sophisticated and, in the process, capable of providing a growing array of services of interest to those endeavoring to solve environmental problems. For the most part, these are specific

---

[14]In 2007 China deliberately destroyed one of its own inoperative meteorological satellites using a ground-based missile, generating more than 3000 long-lived fragments; half of that number of debris fragments were produced by an accidental collision of two US-operated satellites (Brachet 2012).

versions of issues arising in a wide range of settings as the power of information systems increases. Should there be rules regarding rights to privacy on the part of those whose activities are being monitored (e.g., emitters of greenhouse gases) or whose activities play a role in causing environmental problems of public concern (e.g., users of chemical fertilizers and pesticides)? Is it important to be concerned about the potential for corrupt or unethical behavior on the part of either providers or users of Earth observation data? Are there or should there be safeguards regarding the handling of such data in the event that anti-democratic forces achieve growing influence or political turmoil ensues? None of these concerns is unique to the realm of Earth observation environmental data; they arise in many settings. Experience with the integrity of information systems in other settings can be applied to the development of an Earth observation environmental regime complex. But it will be important for both providers and users to be conscious of these concerns and to adapt general responses to the specific circumstances arising in conjunction with Earth observation environmental data.

16.7 Conclusion

We have approached the development of an Earth observation environmental regime complex largely as a matter of building productive relationships between two distinct communities to enhance capacity to provide information, assist in preventing or mitigating disasters, and improve compliance regarding a range of environmental problems. We expect that the provider community, including private companies as well as public space agencies, and the user community, including those responsible for building and implementing a wide range of environmental regimes, will remain distinct. But it will be important to devise and strengthen terms of engagement that allow the two communities to develop end-to-end relationships starting with the design of space missions and payloads and carrying through to the assessment of performance in addressing a variety of tasks and improving the design of future missions. In our judgment, there is room not only for minimizing tensions in the relationships between these communities but also to achieve a higher level of synergy that improves the performance of environmental regimes and solidifies support for space agencies. Making use of our general knowledge of regime interplay, we have identified the principal elements of the terms of engagement between the two communities and discussed ways to develop these terms in a manner that benefits both communities going forward. One practical step that may prove helpful in this context is to launch an Earth observation environmental data forum with a mandate to foster enhanced coordination between the two communities.

References

Adriaensen M et al (2015) Priorities in national space strategies and governance of the member states of the European Space Agency. Acta Astronaut 117:356–367

Bailey G et al (2001) International collaboration: the cornerstone of satellite land remote sensing in the 21st century. Space Policy 17:161–169

Belward A, Skøien JO (2015) Who launched what, when and why; trends in global land-cover observation capacity from civilian Earth observation satellites. ISPRS J Photogrammetry Remote Sens 103:115–128

Biermann F et al (2009) The fragmentation of global governance architectures: a framework for analysis. Glob Environ Politics 9:14–40

Bojinski S et al (2014) The concept of essential climate variables in support of climate research, applications, and policy. Bull Am Meteorol Soc 95:1431–1443

Brachet G (2012) The origins of the "Long-term Sustainability of Outer Space Activities" initiative at UN COPUOS. Space Policy 28:161–165

Calle A, Casanova JL (2008) Forest fires and remote sensing. In: Coskun HG et al (eds) Integration of information for environmental security. Springer, Dordrecht

Cobb, WNW (2011) Who's supporting space activities? An 'issue public' for US space policy. Space Policy 27:234–239

Constable AJ et al (2016) Developing priority variables ("ecosystem Essential Ocean Variables"— eEOVs) for observing dynamics and change in Southern Ocean ecosystems. J Marine Sys 161:26–41

Davies DK et al (2009) Fire Information for Resource Management System: archiving and distributing MODIS active fire data. IEEE Trans Geosci Remote Sens 47(1):72–79

Dillow RK (2012) World Weather Watch. In: Encyclopedia of Global Warming & Climate Change. SAGE Publications, Thousand Oaks

Edwards PN (2006) Meteorology as infrastructural globalism. Osiris 21:229–250

European Commission (2013) Commission Delegated Regulation (EU) No 1159/2013 supplementing Regulation (EU) No 911/2010 of the European Parliament and of the Council on the European Earth monitoring programme (GMES) by establishing registration and licensing conditions for GMES users and defining criteria for restricting access to GMES dedicated data and GMES service information. Official Journal of the European Union L 309/1

European Space Agency (2012) ESA data policy for ERS, Envisat and earth explorer missions. https://earth.esa.int/c/document_library/get_file?folderId=296006&name=DLFE-3602.pdf. Accessed 6 Sep 2016

FAO, Food and Agriculture Organization (2015) Report of the third session of the Joint FAO/IMO ad hoc Working Group on Illegal, Unreported and Unregulated (IUU) Fishing and Related Matters. http://www.fao.org/3/a-i5736e.pdf. Accessed 8 Dec 2016

Gaubert A (2002) Public funding of space activities: a case of semantics and misdirection. Space Policy 18:287–292

GEO, Group on Earth Observations (2015a) GEOSS data management principles. https://www.earthobservations.org/documents/dswg/201504_data_management_principles_long_final.pdf. Accessed 6 Sep 2016

GEO, Group on Earth Observations (2015b) Mexico City Declaration. http://www.earthobservations.org/documents/ministerial/mexico_city/MS3_Mexico_City_Declaration.pdf. Accessed 3 Aug 2016

Handberg R (2003) Reinventing NASA: human spaceflight, bureaucracy, and politics. Praeger, Headport

Harris R, Baumann I (2015) Open data policies and satellite Earth observation. Space Policy 32:44–53

Harris R, Browning R (2003) Global monitoring for environment and security: data policy considerations. Space Policy 19:265–276

Hayes KR et al (2015) Identifying indicators and essential variables for marine ecosystems. Ecol Ind 57:409–419

IEG, World Bank Independent Evaluation Group (2006) Hazards of nature, risks to development: an IEG evaluation of World Bank assistance for natural disasters. World Bank, Washington, D.C., USA

Kuriyama I (2005) Supporting multilateral environmental agreement with satellite Earth observation. Space Policy 21:151–160

Lucas R et al (2014) Contribution of L-band SAR to systematic global mangrove monitoring. Mar Freshw Res 65:589–603

McCauley DJ (2016) Ending hide and seek at sea. Science 351:1148–1150

Morel P (2013) Advancing Earth observation from space: a global challenge. Space Policy 29:175–180

Oberthür S, Gehring T (2006) Institutional interaction in global environmental governance: synergy and conflict among international and EU policies. MIT Press, Cambridge

Oberthür S, Stokke OS (2011) Managing institutional complexity: regime interplay and global environmental change. MIT Press, Cambridge, Massachusetts, USA

OECD (2012) Meeting global challenges through better governance: international co-operation in science, technology and innovation. OECD Publishing, Paris

OECD (2014) The space economy at a glance. OECD Publishing, Paris

Orsini A et al (2013) Regime complexes: a buzz, a boom, or a boost for global governance? Glob Gov 19:27–39

Ouellette W, Getinet W (2016) Remote sensing for marine spatial planning and integrated coastal areas management: achievements, challenges, opportunities and future prospects. Remote Sens Appl Soc Environ 4:138–157

Pereira H et al (2013) Essential biodiversity variables. Science 339:277–278

Raustiala K, Victor DG (2004) The regime complex for plant genetic resources. Int Organ 58(2):277–309

Rosenqvist A et al (2014) Global Mangrove Watch (GMW)—project update and progress in its implementation as a pilot to the Ramsar GWOS. In: Scientific and Technical Review Panel of the Ramsar Convention on Wetlands 18

Reddy S et al (2015) Enhancing the national carbon accounting and reporting capability using remote sensing data. In: Applications of satellite Earth observations: serving society, science, and industry. http://ceos.org/document_management/Publications/Data_Applications_Report/DAR_Summary-Brochure_Digital-Version_Dec2015.pdf. Accessed 8 Dec 2016

Royal Society (2015) Observing the Earth—expert views on environmental observation of the UK. R Soc, London

Sawyer G, de Vries M (2012) About GMES and data: geese and golden eggs—a study on the economic benefits of a free and open data policy for GMES Sentinels data. European Association of Remote Sensing Companies, Brussels

Schwach V (2000) Havet, fisken, og vitenskapen. Fra fiskeriundersøkelser til Havforskningsinstitutt, Bergen

Selding PB (2014) France to make older Spot images available to researchers for free. http://spacenews.com/39234france-to-make-older-spot-images-available-to-researchers-for-free/. Accessed 6 Sep 2016

Sene K (2008) Flood warning, forecasting and emergency response. Springer Science + Business Media, New York

Smith LJ, Doldirina C (2008) Remote sensing: a case for moving space data towards the public good. Space Policy 24:22–32

Stokke OS (2014) Actor configurations and compliance tasks in international environmental governance. In: Kanie N et al (eds) Improving global environmental governance. Routledge, New York

Stokke OS (2015) Fisheries and whaling. In: Pattberg PH, Zelli F (eds) Encyclopedia of global environmental governance and politics. Edward Elgar, Cheltenham

202   O.S. Stokke and O.R. Young

Stokke OS, Oberthür S (2011) Introduction: institutional interaction in global environmental change. In: Oberthür S, Stokke OS (eds) Managing institutional complexity: regime interplay and global environmental change. MIT Press, Cambridge, Massachusetts, USA

Sweta LO, Bijker W (2013) Methodology for assessing the usability of Earth observation-based data for disaster management. Nat Hazards 65:167–199

Tahvonen K (2010) Monitoring oil pollution from ships: experiences from the Northern Baltic Basin. In: Vidas D (ed) Law, technology and science for oceans in globalization: IUU fishing, oil pollution, bioprospecting, outer continental shelf. Martinus Nijhoff, Leiden

Tan SY (2014) Meteorological satellite systems. Springer, New York

UNECE, United Nations Economic Commission for Europe (2009) Transboundary flood risk management: experiences from the UNECE region. United Nations, New York

Viikari L (2007) The environmental element in space law: assessing the present and charting the future. Martinus Nijhoff, Leiden

Weeden BC, Chow T (2012) Taking a common-pool resources approach to space sustainability: a framework and potential policies. Space Policy 28:166–172

Williamson RA (2012) Assuring the sustainability of space activities. Space Policy 28:154–160

Withee GW et al (2004) Progress in multilateral Earth observation cooperation: CEOS, IGOS and the ad hoc Group on Earth Observations. Space Policy 20:37–43

Young OR (1996) Institutional linkages in international society: polar perspectives. Glob Gov 2 (1):1–23

Young OR (2002) The institutional dimensions of environmental change: fit, interplay, and scale. MIT Press, Cambridge

Author Biographies

Olav Schram Stokke is a professor of political science at the University of Oslo, the director of the university's cross-disciplinary bachelor's program on international relations, and a research professor at the Fridtjof Nansen Institute (FNI), where he also served as research director for many years. Previous affiliations include the Centre for Advanced Study (CAS) at the Norwegian Academy of Science and Letters and the International Institute for Applied Systems Analysis (IIASA). His area of expertise is international relations with special emphasis on institutional analysis, resource and environmental management, and regional cooperation. Among his recent books are *Disaggregating International Regimes: A New Approach to Evaluation and Comparison* (MIT Press 2012), *Managing Institutional Complexity: Regime Interplay and Global Environmental Change* (MIT Press 2011), and *International Cooperation and Arctic Governance* (Routledge 2007/2010, Chinese version by Ocean Press of China 2014). He publishes in leading international journals, including *Annals of the American Academy for Political and Social Science, Cooperation and Conflict, Global Environmental Politics, International Environmental Agreements, International Journal of Business Research, Marine Policy, Ocean and Coastal Management, Ocean Development and International Law*, and *Strategic Analysis*.

Oran R. Young is a renowned Arctic expert and a world leader in the fields of international governance and environmental institutions. His scientific work encompasses both basic research focusing on collective choice and social institutions, and applied research dealing with issues pertaining to international environmental governance and the Arctic as an international region. Professor Young served for 6 years as vice-president of the International Arctic Science Committee and was the founding chair of the Committee on the Human Dimensions of Global Change within the National Academy of Sciences in the USA. He currently chairs the Scientific Committee of the International Human Dimensions Programme on Global Environmental Change and the Steering Committee of the Arctic Governance Project.

Chapter 17
Conclusion

Masami Onoda

17.1 Key Findings

In the previous chapters, experts have contributed examples of their work on the impacts of space-based Earth observation on society and policy and have provided insights on how these impacts can be quantitatively assessed. The case studies in Parts II-IV demonstrate how specific countries, organizations, or projects have attempted to link societal needs and global policy demands to the benefits of satellite Earth observations. Chapter 16 examines topics related to a future Earth Observation Regime Complex and a possible Earth Observation Environmental Data Forum.

The points of departure for the book were laid out in the key findings from the Policy and Earth Observation Innovation Cycle (PEOIC) Advisory Board workshop held in Tokyo in November 2015 (see Appendix C for the full text):

1. Earth observations provide a unique window and perspective on our world, serving the betterment of all humankind by supporting policies aimed at sustainably managing natural and societal resources on an ever more populous, affluent, and interconnected planet Earth.
2. Earth observations should be regarded as critical societal infrastructure. There is strong evidence that publicly open Earth observations are making positive, cost-effective contributions to solving a variety of high priority environmental and societal problems.
3. There is a need to develop appropriate institutions in the field of Earth observation through a process to ensure that the observations and prediction systems

M. Onoda (✉)
International Relations and Research Department, Japan Aerospace Exploration Agency,
4-6 Kanda Surugadai, Chiyoda-Ku, 101-8008 Ochanomizu Sola City, Tokyo, Japan
e-mail: onoda.masami@jaxa.jp

© The Author(s) 2017
M. Onoda and O.R. Young (eds.), *Satellite Earth Observations and Their Impact on Society and Policy*, DOI 10.1007/978-981-10-3713-9_17

are comprehensively exploited for policy-making with full engagement of all stakeholders and end users.

4. Japan, together with its international partners, should identify and fill emerging gaps in next generation space missions to guarantee full realization of all societal benefits of Earth observations derived from long-term continuity.

5. There is a changing paradigm for Earth observations, with nongovernmental groups launching satellites and with the growing popularity of small satellites, drones and crowd-sourcing/citizen science campaigns, which are associated with the rapid development of data technology and applications.

Each author contribution builds on these statements and elaborates in detail lessons learned from their individual experiences. All contributions recognize the growing *impact of human action on the Earth system*, the efforts to *integrate natural and social sciences*, and the usefulness of space-based Earth observations that provide a unique viewpoint for understanding this interaction and *supporting decision-making by providing scientific evidence*. A key message from the contributions is the need to demonstrate the *value* or *ROI* of Earth observations. The importance of *open access to data* for increased societal benefit is also stressed. However, authors note the *challenge of quantitatively assessing* the societal and economic impacts of Earth observations, as well as the *need to select the right analytical technique* for the specific context in question. Finally, the impact of *new technologies* is recognized as an opportunity to extract actionable knowledge, as well as a challenge to the assessment of benefits, as they will combine space-based Earth observations with other sources of information.

17.2   Assessment Methodologies

The assessment methodologies introduced in this volume can be categorized largely into literature-based analysis and economic analysis.

17.2.1   Literature-Based Analysis

The PEOIC project of Japan (Chap. 2) takes a unique approach in combining literature analysis with data mining techniques to assess how satellite Earth observation data and information have been used in policy formation and implementation, based on an institutional approach featuring the case of the protection of the stratospheric ozone layer. Satellite Earth observations are a key monitoring tool for the regime, providing decision-makers with information on the status of the environment (i.e., systematic observations of the ozone layer). The study applies a text-mining technique to investigate the relationship between policy and satellite Earth observations by selecting literature that represent policy decisions and scientific reports. The work is significant and unique as there has been no such analysis performed in the past. Constructing comparable literature databases proved challenging.

Presented in Chap. 14, Johnson et al. attempt an assessment of the quantity of Earth observation data used for Reducing Emissions from Deforestation and Forest Degradation in Developing Countries (REDD+) implementation. Results show that Earth observation data *quality* (i.e., spatial resolution) was basically uniform among users, while data *quantity* (i.e., the number of maps produced, the length of historical timeframe for which data were used) varied considerably from country to country and project to project, while neither the data quality nor quantity explained the ways in which the countries or projects utilized the data.

Earlier work on the application of Landsat data (Macauley 2008) followed a similar but distinct approach. Further, the Group on Earth Observations (GEO) conducted an analysis to identify Earth observation priorities for different Societal Benefit Areas (SBAs) and then performed a cross-SBA analysis based on documents from the past 10 years (Group on Earth Observations 2010). As useful as these studies are in providing quantitative evidence of the benefits of Earth observations for addressing policy priorities, literature-based studies have limitations mainly due to the fact that outside the scientific domain, it is rare that a specific data source, satellite, or sensor name will be mentioned in policy literature. The challenge therefore is to identify a meaningful correlation between satellite data and policy decisions, given that satellite data are almost always combined with other data sources before they are processed into information useful to decision-makers. The PEOIC study does succeed in showing a quantifiable relationship between scientific assessments and policy decisions. By establishing the role of satellite data within the scientific assessments it attempts to identify the role of satellite Earth observation in decision-making. Evolving information technology will allow more progress to be made in this area.

17.2.2 Economic Analysis

Economic analysis is a method applied increasingly to measure, account for, and demonstrate the benefits of Earth observations in many societal settings. The need has been prompted in particular by growth of economic constraints in most major space programs. In Chap. 7, Lafaye notes that assessing benefits for public policies and the development of commercial space-based services is now part of the framework of French Earth observation programs.

For the U.S. perspective, Onoda presents Macauley's analysis of rationales and processes for investment in Earth observations to serve national needs (Chap. 3). Emphasis is given to the value of Earth observations in meeting national imperatives in the management of natural and environmental resources, including energy, water, forests, and air quality. The chapter presents an overview of legislation, interagency cooperation, and the Decadal Survey and national assessment processes to inform Earth observation investment. The concept of Value of Information (VOI) is introduced, along with a summary of its applicability as a measure of the economic value of Earth-observing systems, referencing specific examples.

In Chap. 4, Friedl provides an introduction to NASA's Applied Earth Science Program and how missions and applications are linked to societal benefits and environmental policy. He introduces two specific examples—air quality and volcanic ash—for which NASA has conducted studies on the socioeconomic impacts of satellite missions and applications. The study on air quality estimated that satellite data represents a value of around 26 million USD. In the case of volcanic ash, it was estimated that the use of satellite Earth observation data for air traffic management following the 2010 Eyjafjallajökull volcanic eruption in Iceland saved 25–72 million USD. If data had been used from the beginning of the incident, costs estimated at an additional 132 million USD might have been avoided.

ESA's Earth observation strategy and the European Copernicus program are introduced by Aschbacher (Chap. 5). The Copernicus program included an extensive study in its mission-planning phase based on prospective (*ex ante*) costs and benefits. Independent studies for Copernicus have shown that, on average, 1 EUR invested in the Copernicus program leads to an economic benefit of approximately 10 EUR due to better decisions, more efficient policy implementation, and savings due to better preparedness in the case of natural disasters, combined with various other economic potentials such as in job creation and the use of imagery. He also notes that these benefits will only be realized if access to space-based Earth observation data and information is full, free, and open.

Lafaye, in Chap. 7, introduces two examples of collaboration between the French space agency (CNES) and Ministries through a master agreement with the French Ministry of Environment (MEDDE) and the Ministry of Overseas Territories (DGOM). In this frame, an analysis of public policies and the identification of satellite Earth observation contributions were conducted, in areas including the management of natural and environmental resources, as well as maritime environmental surveillance and security. CNES has also initiated studies to assess the economic impact of government investment in the downstream sector of satellite Earth observation. The article provides an informative set of lessons learned and perspectives.

In Chap. 15, Jolly introduces the activities of the OECD Space Forum and examines their methods (both macro- and micro-economic) for assessing socioeconomic benefits derived from space technologies. An important lesson is that existing methods provide useful insights into the socioeconomic benefits derived from space infrastructure, but with a range of caveats. There is a need for further refinement of quantitative analytical tools. She concludes that it remains essential to maintain the effort to build the international knowledge base, an effort that OECD is taking up with the major space-related organizations in OECD economies.

Obersteiner et al. contribute a visionary description of their work to develop methodologies and analytical tools to assess the societal benefits of investments in improving the Global Earth Observation System of Systems (GEOSS) across its nine SBAs (Chap. 12). The fundamental idea of this work is that costs incurred in incrementally improving the observing system will result in benefits through information cost reduction or better-informed decisions. The assessment finds that in the majority of case studies, the societal benefits of improved and globally coordinated Earth observation systems were orders of magnitude higher than the investment costs. It concludes that GEOSS-informed Earth system science products and services must receive adequate support to guarantee a transition to more scientific and evidence-based decision-making.

17.3   Perspectives on the Benefits of Earth Observations for Society and Policy

Sir Martin Sweeting introduces a U.K. perspective in Chap. 6. Observations of the environment are of critical importance to the U.K., providing scientific evidence to support decision-making. There is a need for comprehensive, continuous, and fresh observation of a growing range of parameters to address the various challenges facing the country. Furthermore, he notes the opportunities from new technologies and the avalanche of data and actionable knowledge they will provide. Sustained contributions from all forms and sources—national and international—are needed and there is a clear need for stakeholder dialog in order to maximize the potential of environmental observations for decision-making.

China (Chap. 8) has developed a top-down policy approach with the Ten-year Plan for the China Integrated Earth Observation System (2007) and the Medium- and Long-Term Plan for Development of Science and Technology (2006–2020). Earth observation technologies are now recognized as major public goods that are urgently required in support of various areas. In this chapter, Fan introduces the key components of the Chinese Earth observation program—including the meteorological, oceanic, Earth resources, high-resolution, and environmental protection and Disaster Monitoring Constellation components—as well as the role of the private sector. The Chinese Earth observation program uses a system-of-systems approach and data are distributed free of charge in line with international practice.

Japan's Greenhouse gases Observing SATellite (GOSAT, 'Ibuki') is the world's first satellite dedicated to measuring atmospheric concentrations of carbon dioxide (CO_2) and methane (CH_4) from space. In Chap. 9, Yokota presents an overview of the GOSAT program, its instruments, the measurements it provides, and plans for its successor, GOSAT-2. He also discusses the potential of satellite measurements to support Measurement, Reporting, and Verification (MRV) for multilateral agreements, in particular for carbon inventory verification. Yokota concludes that free and open data and information exchange are essential to realize this potential.

Nakajima, in Chap. 10, introduces the Japanese satellite Earth observation program, and suggests that there will be a gap in Earth observation capacity after 2022. He urges Japan to reinforce its national satellite planning, advocating a bottom-up planning approach. The Dream Roadmap toward 2050, proposed by the

Science Council of Japan is introduced, along with many other requirements for sustained Earth system observations. Japan's Earth observation plans for missions and applications are discussed and the need to enhance Japan's national Earth observation program planning process is reiterated. He further points out the need to connect satellite Earth observation with advanced modeling and societal services.

GEO has established the new decadal strategy of the GEOSS (Chap. 11). Since 2005, GEO has had considerable success in developing GEOSS, advocating broad and open data sharing, initiating major global monitoring initiatives, strengthening these accomplishments, and recognizing the need for further collective effort to foster the use of Earth observation resources to their fullest extent. The GEO Strategic Plan 2016–2025 is introduced in Sect. 4.1. The Plan builds on the strong foundations of GEO, identifying improvements including strengthening the SBAs; engaging more broadly with stakeholders; establishing a robust, steady resourcing mechanism within the voluntary framework of GEO; and identifying new opportunities.

Highlighting a specific case of a Multilateral Environmental Agreement (MEA), Rosenqvist describes in Chap. 13 how Earth observation techniques are used to support the UNFCCC agreement on REDD+. The Global Forest Observations Initiative (GFOI) is presented, and outcomes and implementation challenges are discussed.

Finally, returning to the taxonomy provided in Chap. 1 on the roles that satellite Earth observations can play in addressing environmental concerns (i.e., inform on progress, assist prevention or mitigation, enhance compliance), the practical next steps may involve a project to map Earth observations against the taxonomy in Table 1.1 and to perform analyses of the extent that each role is being addressed, based on the various methodologies presented in the above contributions. Such analyses would help us understand, from an institutional point of view, the role and impact of satellite Earth observations on environmental policy. In Chap. 16, Stokke and Young explore this question in detail and ask whether these capabilities have progressed to the point where there is a compelling case to create an integrated Earth Observation Regime Complex, which will, if realized, complete the innovation cycle from science to policy. They conclude with a discussion of a future international Earth Observation Data Forum with a mandate to foster enhanced coordination between providers and the user community.

17.4 A Model of Earth Observation for Society and Policy, and Lessons for Japan

Based on the contributions to this volume and the 2-year study of the PEOIC project, we can describe a general model as shown in Fig. 17.1.

Starting with stakeholder dialog, the model starts with the identification of taxonomy, segments, actors, the value chain, and values associated with the societal

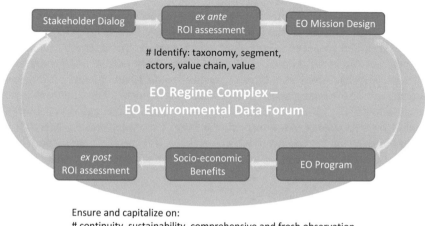

Fig. 17.1 A model of Earth observation for society and policy

or policy goals of the mission, as well as an *ex ante* (prospective) assessment of ROI. The optimal mission design can be established based on this assessment. The program then moves into the development and operation phases. The assessment guarantees that the mission provides comprehensive observations that are adapted to user needs and ensure continuity and sustainability. Sustained funding and stakeholder dialog are critical. Commercial capabilities and new technologies should be leveraged in a timely manner; these companies should be involved as stakeholders throughout the process. Finally, an international knowledge base, such as that of the OECD Space Forum, should be maintained so that experiences can be shared for subsequent missions. The model can be applied to various societal or scientific domains, targeted either to specific nations or international programs; it can also be combined with the concept of an Earth observation regime complex to effectively carry out the programs.

The lack of a systematic approach such as the one described above may have slowed the future planning of the Japanese Earth observation program. Now that Japan's space program has matured, and given the economic constraints Japan has been facing over the last two decades, it seems timely for Japan's space program to take concrete steps toward establishing such an effective scheme, together with international partners and stakeholders. It is time to realize that the focus of space-based Earth observations is no longer purely R&D or science; it must also serve the needs of those who finance it.

The PEOIC project started as an effort to look to the international community for guidance on the development of tools to assess the benefits of Earth observations. With the establishment of the International Advisory Board, the project evolved beyond its initial scope, enabling us not only to compile the latest achievements in the field but also to add a novel view on the future governance of Earth observation systems. This has given us a broader perspective on where we are heading with existing and newly emerging Earth observation technologies. The study was timely because today we are at the point where we have the technological capability to exploit the avalanche of new information and thus knowledge that are derived from Earth observations for improved decision-making. It is therefore our responsibility to use this opportunity well, to achieve a better future for human society and the Earth's environment.

References

Group on Earth Observations (2010) GEO Task US-09-01a: critical Earth observation priorities, final report. https://sbageotask.larc.nasa.gov/Final_SBA_Report_US0901a_Apr2011.pdf. Accessed 8 Dec 2016

Macauley MK (2008) Earth observations in social science research for management of natural resources and the environment: identifying the landsat contribution. J Terr Obs 1(2):31–51

Author Biography

Masami Onoda is currently the U.S. and multilateral relations interface at the International Relations and Research Department of the Japan Aerospace Exploration Agency (JAXA). As an academic, she is fellow of the Institute of Global Environmental Strategies (IGES) and she is also engaged in the private sector as an advisor to the Singapore-based space debris start-up Astroscale Pte. Ltd. since its foundation in 2013. From 2009 to 2012, Dr. Onoda was a scientific and technical officer at the intergovernmental Group on Earth Observations (GEO) Secretariat in Geneva, Switzerland. From 2003 to 2008, while pursuing her graduate studies, she was invited to the JAXA Kansai Satellite Office in Higashiosaka as a space technology coordinator to support technology transfer to SMEs for the small satellite project SOHLA-1. From 1999 to 2003, she worked in the field of Earth observations at JAXA (then NASDA), serving on the Secretariat of the Committee on Earth Observation Satellites (CEOS). In 1999, she was seconded to the UN Office for Outer Space Affairs (UNOOSA) for the organization of the UNISPACE III conference. She holds a Ph.D. in global environmental studies (2009) and a master's degree in environmental management (2005), both from the Kyoto University Graduate School of Global Environmental Studies. Her undergraduate degree is in international relations from The University of Tokyo.

Glossary

ACE Atmospheric Chemistry Experiment
AHI Advanced Himawari Imager
AIC Akaike Information Criterion
AIS Automatic Identification System
AMSU-A Advanced Microwave Sounding Unit-A
APFM Associated Programme on Flood Management
ArCS Arctic Challenge for Sustainability
ARD Analysis-Ready Data
BON Biodiversity Observation Network
BUV Backscatter Ultraviolet
CAI Cloud and Aerosol Imager
CAS Centre for Advanced Study
CBERS China-Brazil Earth Resources Satellite
CCAC Climate and Clean Air Coalition
CCAMLR Convention on the Conservation of Antarctic Marine Living Resources
CCD Charged Coupled Devices
CCOL Coordinating Committee on the Ozone Layer
CEOS Committee on Earth Observation Satellites
CFCS Conservation of Forest Carbon Stocks
CFCs Chlorofluorocarbons
CGI Commissariat Général à l'Investissement
CGMS Coordination Group for Meteorological Satellites
CHEOS China High-resolution Earth Observation System
CLS Collecte Localisation Satellites
CMA China Meteorological Administration
CMS Convention on the Conservation of Migratory Species of Wild Animals
CNES Centre National d'Études Spatiales

© The Author(s) 2017
M. Onoda and O.R. Young (eds.), *Satellite Earth Observations and Their Impact on Society and Policy*, DOI 10.1007/978-981-10-3713-9

COP Conference of the Parties
COPUOS Committee on the Peaceful Uses of Outer Space
COSPACE Le Comité de Concertation Etat Industrie sur l'Espace
DD Future deforestation and/or forest degradation
DEAL Direction de l'Environnement et de l'Aménagement du Littoral
DF Deforestation
df Document frequency
DGE Direction Générale des Entreprises
DGOM Direction Générale des Outre-Mer
DHF Data Handling Facility
DMC Disaster Monitoring Constellation
EARSC European Association of Remote Sensing Companies
EBVs Essential Biodiversity Variables
EC European Commission
ECVs Essential Climate Variables
EMSA European Maritime Safety Agency
EO Earth Observation
EOA Earth Observation Assessment
EOEP Earth Observation Envelope Programme
EORC Earth Observation Research Center
EPA Environmental Protection Agency
ESA European Space Agency
ESF European Science Foundation
ESM Ecosystems Services and Management
EU European Union
EUMETSAT European Organisation for the Exploitation of Meteorological Satellites
FAA Federal Aviation Administration
FAO Food and Agriculture Organization
FCPF Forest Carbon Partnership Facility
FCSE Forest Carbon Stock Enhancement
FCT Forest Carbon Tracking
FD Forest Degradation
FEC Forecast Error Contribution
FFI Flood Forecasting Initiative
FIRMS Fire Information for Resource Management System
FNI Fridtjof Nansen Institute
FRA Forest Resources Assessments
FRELs/FRLs Forest Reference Emission Levels/Forest Reference Levels
FTS Fourier Transform Spectrometer
FY FengYun

G8 Group of Eight Industrialized Nations
GARP Global Atmospheric Research Programme
GBFCI Ground-Based Forest Carbon Inventory
GCOS Global Climate Observing System
GDAS GOSAT Data Archive Service
GDPFS Global Data-Processing and Forecasting System
GEF Global Environment Facility
GEMS Global Environment Monitoring System
GEO Group on Earth Observations
GEOSS Global Earth Observation System of Systems
GF GaoFen
GFCS Global Framework for Climate Services
GFOI Global Forest Observations Initiative
GHG Greenhouse gases
GIFAS Groupement des Industries Françaises Aéronautiques et Spatiales
GIS Geographic Information System
GMES Global Monitoring for Environment and Security
GMW Global Mangrove Watch
GOFC-GOLD Global Observation for Forest Cover and Land Dynamics
GOOS Global Ocean Observing System
GOSAT Greenhouse gases Observing SATellite
GPG Good Practice Guidance
GPM Global Precipitation Measurement
GPS Global Positioning System
GRACE-DAS Gravity Recovery and Climate Experiment-Data Assimilation System
GRENE Green Network of Excellence
GRID Global Resource Information Database
GSMaP Global Satellite Mapping of Precipitation
GUIG GOSAT User Interface Gateway
GWP Global Water Partnership
HABs Harmful Algal Blooms
HBFC Hydrobromofluorocarbons
HFC Hydrofluorocarbon
HIS Hyperspectral Imaging System
HJ HuanJing
HPCI High Performance Computer Infrastructure
HY HaiYang
IAMAS International Association of Meteorology and Atmospheric Sciences
IBAMA Brazilian Institute for the Environment and Renewable Natural Resources
ICES International Council for the Exploration of the Sea

ICSU International Council for Science
Idf Inverse document frequency
IGES Institute for Global Environmental Strategies
IGFA International Group of Funding Agencies for Global Change Research
IIASA International Institute for Applied Systems Analysis
IMO International Maritime Organization
INDC Intended Nationally Determined Contribution
INSPIRE Infrastructure for Spatial Information in the European Community
IOC Intergovernmental Oceanographic Commission
IPBES Intergovernmental Science-Policy Platform on Biodiversity and Ecosystem
          Services
IPCC Intergovernmental Panel on Climate Change
IR Increasing Ratio
IRC International Radiation Commission
IRIS Infrared Interferometer Spectrometer
IRMSS Infrared Multispectral Scanner
ITU International Telecommunications Union
IUU Illegal, Unreported, and Unregulated
JAXA Japan Aerospace Exploration Agency
JICA Japan International Cooperation Agency
JMA Japan Meteorological Agency
JPSS Joint Polar Satellite System
JST Japan Science and Technology Agency
KSAT Kongsberg Satellite Services
L1B Level 1B
L2 Level 2
L3 Level 3
L4A Level 4A
L4B Level 4B
LAI Leaf Area Index
LEO Low Earth Orbit
LIDAR Light Detection and Ranging
LIMS Limb Infrared Monitor of the Stratosphere
LLGHGs Long-Lived Greenhouse Gases
MEA Multilateral Environmental Agreement
MEDDE Ministère de l'Environnement, de l'Énergie et de la Mer
MEXT Ministry of Education, Culture, Sports, Science and Technology
MGD Methods and Guidance Documentation
MLS Microwave Limb Sounder
MOE Ministry of the Environment
MOP Meeting of the Parties

MOPITT Measurements of Pollution In The Troposphere
MRV Measurement, Reporting, and Verification
MST Mesosphere-Stratosphere-Troposphere
NASA National Aeronautics and Space Administration
NASDA National Space Development Agency of Japan
NDC Nationally Determined Contribution
NEAFC North East Atlantic Fisheries Commission
NESDIS National Environmental Satellite, Data, and Information Service
NFMS National Forest Monitoring System
NICFI Norwegian International Climate and Forest Initiative
NICT National Institute of Information and Communications Technology
NIES National Institute for Environmental Studies
NII National Institute for Informatics
NLP Natural Language Processing
NOAA National Oceanic and Atmospheric Administration
NOx Nitrogen oxides
NSMC National Satellite Meteorological Center
OCO-2 Orbiting Carbon Observatory-2
ODS Ozone Depleting Substances
OECD Organisation for Economic Co-operation and Development
OMI Ozone Monitoring Instrument
ONR Office of Naval Research
ORM Ozone Research Managers
PEOIC Policy and Earth Observation Innovation Cycle
PEPS Plateforme d'Exploitation des Produits Sentinel
PIA Programme d'Investissement d'Avenir
PPP Public-Private Partnership
PSI Public Sector Information
RDC R&D Coordination
REDD-MF REDD+ Methodology Framework
REDD+ Reducing Emissions from Deforestation and Forest Degradation in Developing Countries
RFF Resources for the Future
RIMS II Regional Input-Output Modeling System
RISTEX Research Institute for Science and Technology for Society
ROI Return on investment
RS Remote sensing
S-VISSR Stretched Visible and Infrared Spin Scan Radiometer
SAGE Stratospheric Aerosol and Gas Experiment
SAMS Stratospheric and Mesospheric Sounder
SAP Scientific Assessment Panel

SAR Synthetic Aperture Radar
SBA Societal Benefit Area
SBUV Solar Backscatter UltraViolet
SCJ Science Council of Japan
SDCG Space Data Coordination Group
SDGs Sustainable Development Goals
SEM Space Environment Monitor
SFM Sustainable Forest Management
SME Solar Mesospheric Explorer
SMEs Small to medium-sized enterprises
SPC Space Policy Committee
SPDR Special Postdoctoral Researcher
SSC Swedish Space Corporation
SST Sea Surface Temperature
SSTL Surrey Satellite Technology Ltd
SSU Stratospheric Sounding Unit
TANSO Thermal And Near-infrared Sensor for carbon Observation
TES Tropospheric Emission Spectrometer
tf Term Frequency
TOMS Total Ozone Mapping Spectrometer
TOVS TIROS Operational Vertical Sounder
UARS Upper Atmosphere Research Satellite
UN United Nations
UNCBD UN Convention on Biological Diversity
UNCCD UN Convention to Combat Desertification
UNDP UN Development Programme
UNECE UN Economic Commission for Europe
UNEP UN Environment Programme
UNESCO UN Educational, Scientific, and Cultural Organization
UNFCCC UN Framework Convention on Climate Change
UNOOSA UN Office for Outer Space Affairs
USAID U.S. Agency for International Development
USFS U.S. Forest Service
USGCRP U.S. Global Change Research Program
USGEO U.S. Group on Earth Observations
USGS U.S. Geological Survey
VAAC Volcanic Ash Advisory Centers
VCS Verified Carbon Standard
VHR Very-High Resolution
VIRR Visible and Infra-Red Radiometer
VMS Vessel Monitoring System

VOI Value of Information
WCRP World Climate Research Programme
WMO World Meteorological Organization
WVC Wide view CCD camera
WWW World Weather Watch
ZY ZiYuan

Printed in the United States
By Bookmasters